NATIONAL REGISTRY
PARAMEDIC
STUDY QUESTIONS

STEVE ROBERTS

ISBN: 978-1-4834-3650-0 (sc)
ISBN: 978-1-4834-3652-4 (hc)
ISBN: 978-1-4834-3651-7 (e)

Library of Congress Control Number: 2015913129

Lulu Publishing Services rev. date: 08/14/2015

*I would like to dedicate this book
to the following people:*

My maternal grandmother, Ethel Brown. She was the driving force behind me becoming and maintaining my paramedic certification.

My Children, Jackie, Chris and Jon. They're the lifeblood and backbone to this book.

My sister, Andrea, for all her hard work in editing and formatting this book. Without her, this book would not be where it is today.

Jay Marquez, owner of the best EMS training facility in Florida. "First Response Training Group." Thank you.

And to my Dad, Larry. Your constant belief in me throughout my life has played a crucial part in who I am today. Thank you for being there when I needed you the most.

PARAMEDIC STUDY QUESTIONS

Test 1

1. Which of the following is not a national group that is involved in future development of EMS?

    A. National Association of Search and Rescue (NASAR)
    B. National Association of EMS Physicians (NAEMSP)
    C. New York State Ambulance Association
    D. National Association of Emergency Medical Technicians (NAEMT)

2. _____ is defined as a state of complete physical, mental and social well being.

    A. Nutrition
    B. Health
    C. Wellness
    D. Fitness

3. The basics of physical fitness include all of the following except:

    A. Strength, flexibility, and back safety
    B. Drug dependencies and other addictions
    C. Cardiovascular
    D. Nutrition and weight control

4. Which of the following is not a nationally recognized level of EMS Technician?

    A. EMT- Paramedic
    B. First Responder
    C. EMT- Basic
    D. EMT- Cardiac Technician

5. A major side effect of certain types of diuretic therapy is:

    A. Hyperkalemia
    B. Hypokalemia
    C. Hypenatremia
    D. Hypocalcemia

6. Progressively deeper, faster breathing alternating gradually with shallow, slower breathing is called:

    A. Biot's respirations
    B. Kussmaul's respirations
    C. Cheyne-Stokes respirations
    D. Agonal respirations

7. The components of wellness include physical well-being, proper nutrition and:

    A. Compliance with prescribed medications
    B. Mental and emotional health
    C. Vaccinations
    D. Past medical history

8. The drug of choice to treat hyperkalemia in the pre-hospital setting is:

    A. Potassium Hydrochloride
    B. Magnesium Sulfate
    C. Calcium chloride
    D. Atropine

9. The principle that allows the paramedic to function in the field under the auspices and license of a physician is called:

    A. Delegation of authority
    B. Medicum ad litum
    C. Medical Direction
    D. Good Samaritan principle

10. Which patient should be transported immediately, with minimal on-scene care and any attempts at stabilization performed en route to the hospital?

    A. Female, age 45, pulse 132, systolic BP 78
    B. Male, age 54, pulse 98, diastolic PB 80
    C. Female, age 28, systolic BP 96, respiratory rate 18
    D. Male, age 60, pulse 115, respiratory rate 12

11. The most important way for a paramedic to identify that a patient is having an MI is by:

    A. Vital signs
    B. History
    C. EKG reading
    D. Secondary survey

12. Poor diet and nutrition contribute to the development of degenerative diseases that include all the following major killers, except:

    A. Gout
    B. Heart disease
    C. Stroke
    D. Diabetes

13. Signs and symptoms of radiation sickness include:

    A. Severe headache
    B. Excessive thirst
    C. Hair loss
    D. Hearing problems

14. Leaving the non-traumatic cardiac arrest victim, after adequate trial of ALS and BLS has been done, to support the survivors would be an example of a Class _____ guideline.

    A. I
    B. III
    C. IIA
    D. IIB

15. Immediate management for chemical burn injuries is:

    A. Irrigation with copious amounts of water
    B. Applying a neutralizing agent
    C. A dry, sterile dressing
    D. Irrigation with alcohol

16. You suspect that a patient has a complete airway obstruction when she:

    A. Cannot cough
    B. Can exhale with some effort
    C. Cannot swallow
    D. Can only whisper

17. Angina is due to the imbalance in:

    A. Supply and demand for oxygenated blood
    B. Catecholamines in the blood
    C. Oxygen in the blood
    D. Hemoglobin in the blood

18. The upper limit of the normal P-R interval is:

    A. 0.12 second
    B. 0.20 second
    C. 0.10 second
    D. 1.0 second

19. In performing emergency synchronized cardioversion, you would synchronize the electrical shock with which of the following wave forms?

    A. QRS complex
    B. R wave
    C. P wave
    D. P-R interval

20. An early sign of increasing intracranial pressure is:

    A. Bradypnea
    B. Noisy respirations
    C. Tachypnea
    D. None of the above

21. _____ volume is the amount of gas inspired or expired in a minute.

    A. Minute
    B. Adequate
    C. Tidal
    D. Inadequate

22. All of the following are true statements concerning acute mountain sickness except:

    A. The most important part of the treatment is descent to a lower altitude
    B. It usually occurs after rapid ascent to elevations of 5000 feet
    C. Symptoms occur because of decreased oxygen saturation in the blood
    D. Symptoms may include headache, dizziness, nausea, vomiting, and irritability

23. You would be likely to receive an order to administer intravenous thiamine to a patient who appeared to be:

    A. Profoundly intoxicated
    B. In status epilepticus
    C. In metabolic shock
    D. Hyperventilating

24. You have been called by a man who says that his wife is going to commit suicide. Upon arrival, your job will be to do all of the following EXCEPT:

    A. Relay information to the hospital so that the physician can direct disposition
    B. Name the type of depression involved
    C. Interview the patient to clarify the situation
    D. Identify immediate threats to the safety of the patient or others

25. A QRS complex is considered abnormal if it lasts longer that how many seconds?

    A. 0.12 seconds
    B. 0.04 seconds
    C. 0.10 seconds
    D. 0.08 seconds

26. The best was to break a front windshield is:

    A. Hurst tool
    B. Center punch to mid windshield
    C. Fire axe
    D. Center punch to corner of windshield

27. Depression is an example of a:

    A. Organic disease
    B. Psychiatric illness
    C. Psychosis
    D. Mood disorder

28.   Which of the following is not a sign of pure right sided heart failure?

     A.   Pedal edema
     B.   Hepatomegaly
     C.   Pulmonary edema
     D.   Positive hepatojugular reflex

29.   At what rate should the chest be compressed for an adult patient?

     A.   120
     B.   100
     C.   60
     D.   80

30.   The pause following the QRS complex is called the:

     A.   T-T complex
     B.   S-T complex
     C.   T wave
     D.   P-R interval

31.   Your patient has an extremity fracture and has a large area of swelling, which of the following signs should you not elicit?

     A.   Tenderness
     B.   Crepitus
     C.   Ecchymosis
     D.   Swelling

32.   Which of the following routes of administration is fastest?

      A. Sublingual
      B. IM
      C. ET
      D. Subcutaneous

33.   One teaspoon equals:

      A. 1 oz
      B. 5cc
      C. 30cc
      D. 10cc

34.   Anytime a patient has received a high energy impact from the clavicles or superior to the clavicles, the paramedic must consider the possibility of:

      A. Neck injury
      B. Hypovolemia
      C. Low blood sugar
      D. Hypoxia

35.   Injuries to the small intestines may produce:

      A. Peritonitis
      B. Intense blood loss
      C. No clinical significance
      D. Respiratory distress

36. All of the following are types of pneumonia except:

    A. Tuberculosis
    B. AIDS
    C. Aspiration
    D. Myoplasma

37. Which drug can cause users to behave violently and aggressively?

    A. PCP
    B. LSD
    C. Phenobarbital
    D. Elavil

38. Which statement about the pain that accompanies a myocardial infarction is incorrect?

    A. The pain is relieved by sublingual nitroglycerin
    B. Pain due to AMI radiates like anginal pain
    C. Patients often describe the pain as "crushing."
    D. The pain is present only during exertion or stress

39. A mini-drip IV set delivers how many drops per minute of solution?

    A. 60gtts
    B. 10gtts
    C. 30gtts
    D. 15gtts

40. A disease process that destroys the myelin sheath and infects the nerve fibers, impairing nerve function is called:

    A. Epliepsy
    B. Multiple sclerosis
    C. Lou Gehrig's disease
    D. Parkinson's disease

41. The hormone responsible for allowing blood sugar to enter the cells is called:

    A. Insulin
    B. Glucagon
    C. Glucose
    D. Glycogen

42. Which type of radiation is the most serious?

    A. Gamma rays
    B. Delta rays
    C. Alpha particles
    D. Beta particles

43. When do you not give lidocaine?

    A. A PVC's
    B. Bradycardia with PVC's
    C. V-fib
    D. V-tach

44. After administration of epinephrine, what drug may be given to help epinephrine stop an allergic reaction?

    A. Diphenhydramine
    B. Oxygen
    C. Terbutaline
    D. Albuterol

45. Hyperextension of the neck, followed by hyperflexion, is common in:

    A. Lateral impacts
    B. Rear-end impacts
    C. Frontal impacts
    D. Rotational impacts

46. The speed of an anaphylactic reaction depends on the route of exposure and the:

    A. Pre-existing medical conditions
    B. Degree of sensitivity
    C. Level of consciousness
    D. Patient's age

47. Arriving at the scene of a vehicle accident, you find one of the autos in flames with a single occupant in the car. Your first priority after assuring scene safety and resource utilization is to:

    A. Apply a cervical collar
    B. Open the airway
    C. Remove him from the vehicle
    D. Check the pulse

48. If you are asked to set up a staging area for additional responding rescue units, which of the following would be the best location?

    A. Upwind, no visual range, in a building
    B. Downwind, no visual, in a building
    C. Upwind, close visual range, in an open area
    D. Downwind, close visual range, in an open area

49. Which of the following penetrating MOI's has the greatest potential for energy Exchange?

    A. Hanging
    B. Shotgun
    C. High power rifle
    D. Knife

50. Which of the following does not suggest child neglect?

    A. Long-standing skin infections
    B. Multiple insect bites
    C. Extreme malnutrition
    D. Prolonged respiratory infection

51. During a multi-casualty incident, a conscious patient presents with a fractured femur, a palpable radial pulse, and a respiratory rate of 24/min. According to START, this patient would be placed into what triage category?

    A. Deceased- black
    B. Delayed- yellow
    C. Minor- green
    D. Immediate- red

52. Which of the following would not be an appropriate treatment for
     a rape victim:

     A. Cleanse the perineum with sterile saline
     B. Assess and treat major trauma
     C. Allow family members to be present
     D. Provide strong emotional support

53. The P wave on an EKG strip reflects what event inside the heart?

     A. Ventricular repolarization
     B. Atrial repolarization
     C. Atrial depolarization
     D. Ventricular depolarization

54. A drug used to decrease ICP is:

     A. Pitocin
     B. Mannitol
     C. Magnesium sulfate
     D. Valium

55. Cushing's syndrome is caused by the hypersecretion of glucocorticoids
     by the _____ glands (s).

     A. Pancreas
     B. Thymus
     C. Adrenal
     D. Reproductive 100mg

56. 100ml of a 1:000 solution will contain what weight of the drug?

     A. 100mg
     B. 10gm
     C. 10mg
     D. 1gm

57. Which of the following is NOT a component of the START triage method?

     A. Neuro-muscular function
     B. Circulation assessment
     C. Mentation-level of consciousness
     D. Respiration assessment

58. Medications and drugs are most often delivered to a newborn through the use of which circulatory vessel?

     A. Jugular vein
     B. Umbilical vein
     C. Umbilical artery
     D. Femoral artery

59. When involved with a motorcycle accident, if the victim was wearing a helmet:

     A. The likelihood of deceleration injury is greatly diminished
     B. The likelihood of head injury is greatly diminished
     C. The likelihood of spinal injury is greatly diminished
     D. Both A and B

60. Which of the meninges is the highly vascular covering of the spinal cord and brain?

   A. Pia mater
   B. Falx cerebelli
   C. Dura mater
   D. Arachnoid membrane

61. Two common tricyclic antidepressants are:

   A. Valium and Thorazine
   B. Compazine and Thorazine
   C. Elavil and Tofranil
   D. Prozac and Zanax

62. What is the paramedic's primary goal in cases of suspected child abuse?

   A. Make sure that the child receives necessary treatment
   B. Ensure that the child is removed from family custody
   C. Gather up any physical evidence to take to the hospital
   D. Ensure the abuser is arrested upon arrival at the hospital

63. Isuprel is a (n):

   A. Parasympathetic blocker
   B. Alpha stimulator
   C. Beta blocker
   D. Beta stimulator

64. What are the classic symptoms of narcotic overdose?

   A. Excitability, hyperactivity, hypertension
   B. Respiratory depression and constricted pupils
   C. Cardiac dysrhythmias and altered mental status
   D. Altered mental status, euphoria, and dilated pupils

65. All of the following are examples of stable patients, except:

   A. Minor illness
   B. Significant MOI with neck injury
   C. Low-grade fever
   D. Minor isolated injury

66. How should foreign bodies in the eye be removed?

   A. With a Q-tip swab
   B. With small splinter forceps
   C. Irrigation with normal saline
   D. Any of the above

67. Anthrax is classified as a:

   A. Enterotoxin
   B. Bacterial infection
   C. Viral infection
   D. Man made organism

68. Diabetes mellitus is a disease characterized by a lack of:

   A. Insulin
   B. Testosterone
   C. Potassium
   D. Epinephrine

69. All of the following are possible causes of precipitating acute pulmonary edema except:

    A. Dysrhythmias
    B. Acute MI
    C. Strep throat
    D. Acute endocarditis

70. A _____ is the displacement of a bone end from its articular surface:

    A. Dislocation
    B. Strain
    C. Fracture
    D. Sprain

71. A focused history is the chronologic history of the patient's:

    A. Family history
    B. Surgeries
    C. Present illness
    D. Medication use

72. A patient in Ventricular tachycardia, unconscious and no vital signs, you should first:

    A. Start CPR
    B. Defibrillate
    C. Give epinephrine down the ET tube
    D. Start an IV and give Lidocaine

73. One of the best ways for EMS personnel to deal with job related stress is to:

    A. Discuss the situation with co-workers
    B. Take sleeping pills at night as needed
    C. Take time away from family and friends
    D. Eliminate all physical exercises

74. The best protection for a Paramedic is:

    A. Proper equipment
    B. Malpractice insurance
    C. Proper training
    D. Good charting

75. The normal gestational period is:

    A. 30 weeks
    B. 40 weeks
    C. 50 weeks
    D. 20 weeks

76. Drug induced physiologic changes on a body function or process are know as a:

    A. Half life
    B. Drug action
    C. Idiosyncrasy
    D. Side effect

77. Which of the following medications is commonly used to treat patients who are victims of organophosphate poisoning?

    A. Atropine sulfate
    B. Flumanzenil
    C. Adenosine
    D. Calcium chloride

78. 220 lbs equals:

    A. 150 kg
    B. 100 kg
    C. 110 kg
    D. 50 kg

79. A burn wound that blisters is an example of:

    A. Second degree burn
    B. First degree burn
    C. Chemical burn
    D. Third degree burn

80. The pressure of pushing water out of the capillary into the interstitial space is referred to as the:

    A. Tissue hydrostatic pressure
    B. Capillary hydrostatic pressure
    C. Tissue colloidal osmotic pressure
    D. Capillary colloidal osmotic pressure

81. Which of the following is an early finding in a crushing injury?

    A. Paresthesia
    B. Paralysis
    C. Paresis
    D. Pulselessness

82. Digitalis, morphine sulfate and atropine are examples of medications originally derived from:

    A. Synthetic sources
    B. Vegetable sources
    C. Animal sources
    D. Mineral sources

83. The Paramedic should check motor and sensory function in all extremities and document their findings before extricating a patient from a vehicle accident:

    A. True
    B. False

84. Normal capillary refill time is usually:

    A. Less than 4 seconds
    B. Less than 1 second
    C. Less than 2 seconds
    D. Greater that 2 seconds

85.   Full dilation of the cervix signifies:

      A.   The end of the third stage of labor
      B.   The beginning of the first stage of labor
      C.   The end of the first stage of labor
      D.   The end of the second stage of labor

86.   What is the correct needle size and insertion angle for an intramuscular
      injection?

      A.   21 gauge at 90 degrees
      B.   25 gauge at 90 degrees
      C.   21 gauge at 45 degrees
      D.   25 gauge at 45 degrees

87.   The description of a drug using molecular and chemical composition
      is the:

      A.   Official name
      B.   Chemical name
      C.   Trade name
      D.   Brand name

88.   The American Medical Association's Code of Medical Ethics, states
      that patients have the right to:

      A.   Make decisions on health care
      B.   Refuse payment when they are dissatisfied with care
           rendered
      C.   Name the physician who will not do their surgery
      D.   Waive their ethical options

89. All pediatric patients who have had seizures should be:

    A. Given acetaminophen to correct fever
    B. Transported to a hospital for evaluation
    C. Given diazepam rectally, IV or IM
    D. Evaluated for signs of abuse or neglect/

90. Cardiac output equals:

    A. Rate x stroke volume
    B. Stroke volume X blood pressure
    C. Rate plus stroke volume
    D. Stroke volume plus systolic blood pressure

91. Front windshields are made from:

    A. Laminated-tempered glass
    B. Laminated safety glass
    C. Tempered glass
    D. Plexi-glass

92. Dilated pupils means:

    A. Cerebral hypoxia
    B. Morphine overdose
    C. Stimulation of the parasympathetic nervous system
    D. Hypoxemia

93. During a fire, your partner falls down a flight of stairs. What method would you use to remove him from the building?

    A. Extremities carry
    B. Clothes drag
    C. Fireman's drag
    D. Fireman's carry

94. If a Paramedic was found to have administered an excessive dose of lasix to a drug addict in order to get "pleasure" out of watching the patient urinate all over themselves, this is an example of:

    A. Gross negligence
    B. Commission of duty
    C. Proximate cause
    D. Abandonment

95. The main reason an increase in intracranial volume results in a significant increase in intracranial pressure is:

    A. Because the skull is a rigid, closed space
    B. The volume of cerebrospinal fluid
    C. A change in the brains oncotic pressure
    D. The cerebral collateral blood supply

96. The most common site of injury in auto accidents involving unrestrained occupants is:

    A. Long bones
    B. Head
    C. Abdomen
    D. Thorax

97. Of the following, which fracture has the highest possibility of being a life threat?

    A. Femur
    B. Humerus
    C. Clavical
    D. Tibia

98. All of the following are possible causes of coma except:

    A. Antibiotics
    B. Head trauma
    C. Hypoglycemia
    D. Drug overdose

99. Which of the following BEST describes a colloid solution?

    A. A solution containing only salt
    B. A balanced electrolyte solution
    C. A solution containing protein
    D. A solution only glucose

100. Off-line medical control primarily relies on:

    A. Protocols
    B. Stop lines
    C. ACLS algorithms
    D. Advanced directives

101. The first drug given after oxygen in an asystolic patient is:

    A. Isuprel
    B. Lidocaine
    C. Epinephrine
    D. Atropine

102. Your patient has a chemical burn to his face and eyes, How should you treat this?

    A. Flush the area with clean cool water prior to and during transport
    B. Apply a paste made of baking soda and alcohol to the eyes
    C. Cover the eyes and face with a dry sterile dressing
    D. Apply a neutralizing agent to counteract any chemical reaction

103. Pediatric patients should double their weight by:

    A. 1 year
    B. 9 months
    C. 6 months
    D. 1 month

104. A neat and professional appearance, well groomed and clean, are important and:

    A. Help to instill confidence in the patient and their families
    B. Very subjective qualities of an EMT-P
    C. An indication of your competency and skills
    D. Show respect to the patient and their family

105. All of the following are signs and symptoms of a hypoglycemic reaction except:

    A. Nervousness, drowsiness, and coma
    B. Frequent urination, increased thirst, and increased appetite
    C. Confusion, irritability, and combative behavior
    D. Rapid pulse, and cold and clammy skin

106. All of the following are known complications of dialysis except:

    A. Hypotension
    B. Stomach ulcers
    C. Disequilibrium syndrome
    D. Chest pain or dysrhythmia

107. Crowning is defined as:

    A. Bulging of the vaginal opening
    B. An engorgement of the placenta
    C. A bearing down pressure on the rectum
    D. Maximum expansion of the mother's belly

108. Goals for EMS public education for the future include:

    A. Exploring and evaluating public education alternatives
    B. Relying on HMO's to get the message out
    C. National core contents to replace EMS program curriculs
    D. All of the above

109.   No breath sounds in one lung field may indicate which of the following conditions?

   A.   Pulmonary embolism
   B.   Partial airway obstruction
   C.   Pneumothorax
   D.   Flail chest

110.   A greenstick fracture is one that is:

   A.   Comminuted
   B.   Open
   C.   Partial
   D.   Impacted

111.   An important fact to remember when bandaging eye injuries is:

   A.   Never apply pressure directly to the eye ball
   B.   Chemical burns to the eyes should not be irrigated
   C.   If only one eye is injured, do not bandage both eyes
   D.   Remove impaled objects rather than stabilizing them

112.   The first civilian ambulance service was started in _____ in 1865.

   A.   Chicago
   B.   New York City
   C.   Cincinnati
   D.   Philadelphia

113. A child involved in a pedestrian -vs- car collision, the injury is:

    A. Likely to be lateral
    B. Unlikely to involve one portion of the body over another
    C. Likely to be posterior
    D. Likely to be frontal

114. All of the following are indications for the use of transcutaneous pacing except:

    A. Symptomatic atrial fibrillation with a slow ventricular rate
    B. Third degree heart block with a slow ventricular rate
    C. Symptomatic sinus bradycardia
    D. Ventricular fibrillation

115. All of the following are types of aneurysms except:

    A. Atherosclerotic
    B. Congenital
    C. Traumatic
    D. Embolic

116. Open fractures may include all but one of the following:

    A. Severe angulation and edema
    B. Spurting blood
    C. Embedded with dirt or impaled objects
    D. All of the above

117.   When considering the risk assessment for cardiovascular disease, the paramedic should take into account all of the following except:

    A.   Exposure to the sun
    B.   Blood pressure
    C.   Triglycerides
    D.   Stress

118.   Plasma loss in a burn patient causes:

    A.   Fluid overload
    B.   Dehydration
    C.   Hemorrhagic shock
    D.   Hypovolemic shock

119.   What is the most commonly used drug in the pre-hospital setting for patients with asthma?

    A.   IM or IV terbutaline
    B.   IV or IM corticosteroid
    C.   Nebulized or SC epinephrine
    D.   Inhaled or nebulized albuterol

120.   Which item below is not evaluated with an APGAR score?

    A.   Pupils
    B.   Appearance and activity
    C.   Grimace or crying
    D.   Respiratory effort

121. In many cases making a precise diagnosis in the field is difficult:
      Therefore, the Paramedic should strive to recognize emergent signs
      and symptoms and then:

      A. Contact medical control and transport
      B. Begin transport and initiate treatment enroute
      C. Transport to the nearest facility
      D. Stabilize and transport

122. Skin color changes in a dark skinned person can be observed in the:

      A. Mouth
      B. Feet
      C. Axilla
      D. Ears

123. The most common MOI in children is:

      A. Bicycle accidents
      B. Falls
      C. Auto-vs-pedestrian collisions
      D. MVC

124. In geriatric patients, which is not normal:

      A. Central cyanosis
      B. Basilar rales
      C. Pedal edema
      D. Poor skin turgor

ANSWER SHEET 1

1.   C – New York State Ambulance Association
2.   B – Health
3.   B – Drug dependencies and other addictions
4.   D – EMT – Cardiac Technician
5.   B – Hypokalemia
6.   C – Cheyne-Stokes respirations
7.   B – Mental and emotional health
8.   C – Calcium chloride
9.   C – Medical Direction
10. A – Female, age 45, pulse 132, systolic BP 78
11. C – EKG reading
12. A – Gout
13. C – Hair loss
14. C – IIA
15. A – Irrigation with copious amounts of water
16. A – Cannot cough
17. A – Supply and demand for oxygenated blood
18. B – 0.20 second
19. B – P – R interval
20. D – None of the above
21. A – Minute
22. B – It usually occurs after rapid ascent to elevations of 5000 ft.
23. A – Profoundly intoxicated
24. B – Name the type of depression involved
25. A – 0.12 seconds
26. C – Fire axe
27. D – Mood disorder
28. C – Pulmonary edema
29. B – 100
30. B - S – T complex
31. B - Crepitus

32. C – ET
33. B – 5cc
34. A – Neck injury
35. A – Peritonitis
36. B – AIDS
37. A – PCP
38. A – The pain is relieved by sublingual nitroglycerin
39. A – 60 gtts
40. B – Multiple sclerosis
41. A – Insulin
42. A – Gamma rays
43. B – Bradycardia with PVC's
44. A – Diphenhydramine
45. B – Rear-end impacts
46. B – Degree of sensitivity
47. C - Remove him from the vehicle
48. A – Upwind, no visual range, in a building
49. C – High power rifle
50. B – Multiple insect bites

ANSWER SHEET 1

51. B – Delayed – yellow
52. A – Cleanse the perineum with sterile saling
53. C – Atrial depolarization
54. B – Mannitol
55. C – Adrenal
56. A – 100 mg
57. A – Neuro – muscular function
58. B – Umbilical vein
59. B – The likelihood of head injury is greatly diminished
60. A – Pia mater
61. C – Elavil and Tofranil
62. A – Make sure that the child receives necessary treatment
63. C – Beta blocker
64. B – Respiratory depression and constricted pupils
65. B – Significant MOI with neck injury
66. C – Irrigation with normal saline
67. B – Bacterial infection
68. A – Insulin
69. C – Strep throat
70. A – Dislocation
71. C – Present illness
72. A – Start CPR
73. A – Discuss the situation with co-workers
74. C – Proper training
75. B – 40 weeks
76. B – Drug action
77. A – Atropine sulfate
78. B – 100 kg
79. A – Second degree burn
80. B – Capillary hydrostatic pressure
81. A – Paresthesia

82. B – Vegetable sources
83. A – True
84. C – Less than 2 seconds
85. C – The end of the first stage of labor
86. A – 21 gauge at 90 degrees
87. B – Chemical name
88. A – Make decisions on health care
89. B – Transported to a hospital for evaluation
90. A – Rate x stroke volume
91. B - Laminated safety glass
92. A – Cerebral hypoxia
93. B - Clothes drag
94. A – Gross negligence
95. A – Because the skull is a rigid, closed space
96. B – Head
97. A – Femur
98. A - Antibotics
99. C – A solution containing protein
100. A - Protocols

ANSWER SHEET 1

101. C- Epinephrine
102. A- Flush the area with clean cool water prior to and during transport
103. C- 6 months
104. A- Help to instill confidence in the patient and their families
105. A- Nervousness, drowsiness, and coma
106. B- Stomach ulcers
107. A- Bulging of the vaginal opening
108. A- Exploring and evaluating public education alternatives
109. C- Pneumothorax
110. C- Partial
111. A- Never apply pressure directly to the eye ball
112. C- Cincinnati
113. A- Likely to be lateral
114. D- Ventricular fibrillation
115. D- Embolic
116. D- All of the above
117. A- Exposure to the sun
118. D- Hypovolemic shock
119. D- Inhaled or nebulized albuterol
120. A- Pupils
121. D- Stabilize and transport
122. A- Mouth
123. B- Falls
124. A- Central cyanosis

Test 2

1. What do orthostatic vital sign changes suggest for a patient with acute abdominal pain?

    A. The patient is a diabetic
    B. The patient has peritonitis
    C. The patient is hypovolemic
    D. The patient has appendicitis

2. Mr. Roberts is so frightened of height that he will not ride in elevators. This fear is called a:

    A. Phobia
    B. Anxiety
    C. Behavior problem
    D. Coping mechanism

3. Dependence is a _____ craving for a chemical agent, resulting from abuse or addiction.

    A. Spiritual
    B. Physical
    C. Psychologic
    D. Neural

4.   What is the most common route of poisoning:

     A. Ingestion
     B. Absorption
     C. Inhalation
     D. Injection

5.   If you have 200mg of Dopamine to add to 250cc of D5W, what will
     be the concentration of Dopamine?

     A. 0.8 mg/ml
     B. 125 mcg/ml
     C. 1.25 mg/ml
     D. 80 mcg/ml

6.   All of the following are leukocyte (WBC) disorders except:

     A. Leukopenia
     B. Acute leukemia
     C. Anemia
     D. Chronic leukemia

7.   Of the following examples, which is not a scenario in which air medical
     transport should be considered?

     A. Lengthy extrication times
     B. Lengthy manual transport out of a remote area
     C. Spinal injury
     D. Lengthy ground transport time

8. A vitamin that plays a key role in carbohydrate metabolism is called:

    A. Calcium
    B. Vitamin E
    C. Thiamine
    D. Niacin

9. A related disease or condition that can result in rapid and total airway obstruction is

    A. Laryngitis
    B. Bronchiolitis
    C. Bronchitis
    D. Epiglottitis

10. Where would you expect to see a wound that is described as distal to the knee?

    A. Hip
    B. Thigh
    C. Calf
    D. Thigh

11. A patient experiencing heat stroke should have his or her temperature rapidly lowered to a target temperature of:

    A. 98 degrees
    B. 110 degrees
    C. 102 degrees
    D. 98.6 degrees

12. The normal range of hematocrit for women is:

    A. 36-46
    B. 55-60
    C. 44-49
    D. 50-54

13. _____ is the number one cause of death in children over one
    year of age:

    A. Trauma
    B. SIDS
    C. Infection
    D. Abuse

14. The factor common to all forms of shock is:

    A. Inadequate tissue perfusion
    B. Decreased blood volume
    C. Rapid pulse rate
    D. Increased peripheral resistance

15. During delivery, you notice that the amniotic fluid is discolored and
    has a foul odor, what should you do first?

    A. Provide five back blows and then five chest thrusts
    B. Intubate the child and give positive pressure ventilations.
    C. Suction the upper airway using a meconium aspirator
    D. Dry, warm, position, suction, and stimulate the child to
       breathe.

16. Order: Give 25 grams of Dextrose from a 100ml ampule of 50% Dextrose. How many ml's Dextrose will be given?

    A. 100 ml's
    B. 50 ml's
    C. 75 ml's
    D. 25 ml's

17. To perform a needle cricothyrotomy, the patient should be placed:

    A. Supine with head and neck in a slightly flexed position
    B. Supine with head and neck hyperextended
    C. In the lateral recumbent position with head and neck hyperextended
    D. In the lateral recumbent position with head and neck in neutral position.

18. Stable angina normally occurs:

    A. During eating
    B. During rest
    C. During sleep
    D. During exercise or stress

19. The brainstem contains all of the following structures except:

    A. Pons
    B. Lowbrain
    C. Medulla
    D. Oblongata

20. Surface exposure to _____ requires removal by gentle brushing prior to copious amounts of water:

    A. Dry lime
    B. Sodium
    C. Phenol
    D. All of the above

21. The T wave on an EKG tracing represents which of the following events?

    A. Repolarization of the atria
    B. Repolarization of the ventriles
    C. Depolarization of the ventricles
    D. Depolorization of the atria

22. When you are caring for more than one trauma patient at a time, you should change your gloves how often?

    A. Whenever they become soiled
    B. For each new procedure
    C. Whenever time permits
    D. For each new patient

23. Which of the following home health services is not a skilled service?

    A. Nursing
    B. Case management
    C. Shopping service
    D. Infant care

24.   A function of the nervous system is to:

      A.   Monitor internal changes of the body
      B.   Provide support for the body structures
      C.   Circulate nutrients to all body cells
      D.   Regulate the neuron

25.   Which of the following signs is NOT found in salicylate ingestion?

      A.   Bradycardia
      B.   Fever
      C.   Vomiting
      D.   Hyperpnea
      E.   Metabolic acidosis

26.   Early signs of Lidocaine toxicity are:

      A.   Tremors and parasthesia
      B.   Seizures and hypertension
      C.   Headache and seizures
      D.   V-tach and tremors

27.   Which of the following conditions best suggests respiratory failure?

      A.   Tachycardia (>130)
      B.   Diaphoresis
      C.   Change in mental status
      D.   Loud, audible stridor

28. The definitive treatment of cardiac tamponade is to:

    A. Administer IV antibiotic therapy
    B. Relieve cardiac compression
    C. Place a chest tube
    D. Decompress the chest

29. What is your first priority?

    A. Separate the walking wounded from the most severely injured.
    B. Establish the morgue area away from the view of the patients.
    C. Set up the triage area close to the treatment and transport areas.
    D. Begin rapid treatment of the most seriously injured patients.

30. Acid-based balance refers to the concentration of:

    A. Hydrogen ions in body fluids
    B. The chief extracellular ion of body fluids
    C. The chief intracellular ion of body fluids
    D. All of the above.

31. Adrenergic receptors are located:

    A. Membranes on the different organs in the body
    B. On the post synaptic nerve fibers
    C. Both are correct
    D. Neither are correct

157. You suspect that your patient has a complete airway obstruction when:

      E. A. Can exhale with some effort
      F. B. Cannot swallow
      G. C. Cannot cough
      H. D. Can only whisper

32. Key aspects of an ongoing assessment include:

      A. Obtaining a baseline EKG
      B. Trending mental status and vital signs
      C. Starting an IV
      D. Observing MOI

33. What condition is the pathophysiological result of a near drowning in sea water?

      A. Pulmonary edema
      B. Metabolic alkalosis
      C. Ventricular fibrillation
      D. Pulmonary embolism

34. An MVA victim cannot move her lower extremities and is numb at the area of the umbilicus down. You should suspect a spinal injury in the area of:

      A. T-10
      B. C-1
      C. T-1
      D. C-5

35. When a paramedic is confronted with a patient who has a serious
     medical emergency and refuses to allow the paramedics to treat her,
     the paramedic should first try:

     A. Trick the patient into being treated in a round about way:
     B. Explain to the patient the seriousness of her condition
     C. Restrain the patient and treat her
     D. Have a bystander convince the patient to be treated

36. Narcan is used to reverse the respiratory depression of:

     A. Cocaine
     B. Darvon
     C. Librium
     D. Valium

37. The wheezing associated with left-sided heart failure result from:

     A. Chest muscle tightness
     B. Fluid in the lungs
     C. Chest wall expansion
     D. Chronic bronchitis

38. The pain of stable angina is brought on by:

     A. Imminent AMI
     B. Difficulty breathing
     C. Exercise or stress
     D. Overuse of nitroglycerin

39.   What is uticaria?

    A.   Hives
    B.   Sneezing
    C.   Rash
    D.   Snoring respirations

40.   What are the signs of circulatory overload in a patient who is receiving IV fluids?

    A.   Falling blood pressure
    B.   Trauma Score lower than 10
    C.   Dyspnea, rales, and rhonchi
    D.   Agitation and clammy skin

41.   The most common conditions for false capnography reading, include all of the following, except:

    A.   Ingestion of fruits prior to obtaining a reading
    B.   Non-perfusing patients
    C.   Presence of a pulmonary embolism
    D.   Premature neonates

42.   The intrathoracic abdomen includes all of the following except:

    A.   Kidneys
    B.   Stomach
    C.   Spleen
    D.   Liver

43. The painless loss of sight in one eye is most generally associated with:

    A. Myopia
    B. Hyphema
    C. Retinal artery occlusion
    D. Stye

44. Oral medication ingestion is taken through the mouth and absorbed where?

    A. The lower gastrointestinal tract
    B. The sigmoid colon
    C. The stomach
    D. The esophagus

45. Your patient is a 38 year old female who is found unconscious on the living room floor. She is not breathing, has pinpoint pupils, and has a fresh puncture wound to her left forearm. She has multiple scars that form a bluish streak over the veins on the back of both hands. This patient is most likely suffering from which of the following?

    A. A narcotic overdose
    B. Anaphylactic shock
    C. A seizure disorder
    D. Multiple spider bites

46. A collection of blood caused by a ruptured blood vessel that is enclosed in an organ, tissue or body space is a/an:

    A. Stroke
    B. Abscess
    C. Hematoma
    D. Aneurysm

47. What is the primary drug used for the management of acute
     anaphylaxis?

     A. Epinephrine
     B. Diphenhydramine HCL
     C. Terbutaline
     D. Methylprednisolone

48. Which size ET tube is appropriate for female or male patient in an
     emergency situation?

     A. 6.0
     B. 8.5
     C. 7.5
     D. 9.0

49. Radio frequencies are measured in:

     A. Megahertz
     B. Milligrams
     C. Watts
     D. Kilograms

50. Which of the following items is NOT standard equipment for
     performing an endotracheal intubation?

     A. 60cc syringe
     B. Suction equipment
     C. Bite block
     D. Laryngoscope

51. The cornerstone of being an effective paramedic is having the ability to:

    A. Think and work under pressure
    B. Perform the ongoing examination on a bumpy highway
    C. Avoid conflict with co-workers
    D. Drive all types of emergency vehicles

52. All of the following are signs of dehydration in infants except:

    A. Dry mucus membranes
    B. Cyanosis
    C. Sunken fontanelles
    D. Poor skin turgor

53. You are assessing a neonate who has a pink body and blue extremities, a pulse rate of 90, positive grimace response, active motion, and irregular respiratory efforts. What is the APGAR score for this infant?

    A. 6
    B. 10
    C. 4
    D. 8

54. The major advantage to the use of BiPAP and CPAP devices is that:

    A. It does not require continuing education
    B. There is no wasting of oxygen
    C. Intubation
    D. Intubation may be avoided where it may have previously been required

55. The care provided in the first few minutes of a life-threatening emergency is called:

     A. Basic life support
     B. CPR
     C. Ongoing assessment
     D. ACLS

56. All of the following are the most common conditions associated with the cardiac arrest patient except:

     A. PEA
     B. Asystole
     C. Ventricular fibrillation
     D. First degree heart block

57. What is the first treatment for a symptomatic patient with atrial fibrillation with a sustained ventricular heart rate greater than 150 beats per minute?

     A. Immediate cardioversion
     B. Administration of lidocaine
     C. Vagal or valsalva maneuvers
     D. Administration of adenosine

58. The maximum allowable time for an intubation attempt is:

     A. 0 seconds
     B. 15 seconds
     C. 5 seconds
     D. 10 seconds

59. The lumbar spine consists of:

     A. 12 vertebrae
     B. 3 to 5 vertebrae
     C. 7 vertebrae
     D. 4 vertebrae

60. What is the correct field treatment for a frost bitten body part?

     A. Cover the frozen part tightly in wet occlusive dressings
     B. Transport the patient to the hospital
     C. Rub the affected part with crushed snow or ice until warmed
     D. Warm the affected part in water maintained at 101 – 106 F before transporting

61. In performing emergency synchronized cardioversion, you should synchronize the electrical shock with which of the following wave forms?

     A. R wave
     B. QRS complex
     C. P wave
     D. P – R interval

62. When does the third stage of labor begin?

     A. As the placenta is delivered
     B. When the cervix is fully dilated
     C. Immediately upon the birth of the baby
     D. When contractions are five minutes apart

63. Personality or emotional stress are often triggered by any of the following, except:

    A. Personal expectations
    B. Feelings of incompetence
    C. Nausea or vomiting
    D. Feelings of guilt

64. The _____ links the endocrine and nervous systems, as well as the mind (psyche) and body.

    A. Hypothalamus
    B. Thalamus
    C. Pineal body
    D. Insula

65. A home monitoring device which detects changes in thoracic or abdominal movement and heart rate is called a:

    A. Pulmonary function meter
    B. Nebulizer
    C. Apnea monitor
    D. Ventilator

66. The term ethics refers to:

    A. Moral code of conduct
    B. Rules, standards, and morals
    C. Professional standards of care
    D. Upgrading standards of care

67. A patient with increased ICP, you would expect:

    A. Decreased blood pressure and decreased pulse
    B. Increased blood pressure and increased pulse
    C. Decreased blood pressure and decreased pulse
    D. Decreased blood pressure and increased pulse

68. Adenocard and Verapamil are both used for:

    A. SVT
    B. Torsades
    C. Calcium channel blocker overdose
    D. Ventricular tachycardia

69. The universal donor blood type is:

    A. O-
    B. A+
    C. B+
    D. O+

70. Over inflating the pilot balloon in an endotracheal tube can cause:

    A. Ischemia of the trachea wall
    B. Damage to teeth and gums
    C. Displacement of the tube
    D. Return of the gag reflex

71. If upon arrival at the scene a car accident, the vehicle is on its side, off the road on an embankment, it will be necessary to rapidly:

    A. Extricate the patient
    B. Gain access
    C. Stabilize the vehicle
    D. Disentangle the patient

72. A helicopter requires a landing zone of approximately _____ feet on relatively level ground:

    A. 100 x 100
    B. 100 x 200
    C. 50 x 50
    D. 50 x 100

73. What does the term stridor refer to?

    A. A high pitched sound upon inspiration from an airway obstruction
    B. A rattling sound associated with fluid in the upper airway
    C. A gurgling sound resulting from fluid in the lower airways
    D. A whistling sound heard upon expiration in asthma patients

74. Which of the following is the correct number of women beaten by a spouse of partner each year in the United States?

    A. 1.5 million
    B. 4 million
    C. 250,000
    D. 750,000

75. Care for the patient who has just experienced an abortion should include all of following except:

    A. Emotional support
    B. High flow oxygen
    C. 2 mg of morphine sulfate
    D. IV of normal saline or lactated ringer's solution

76. For the treatment of a venomous snakebite, the paramedic should do all of the following except:

    A. Apply a constricting band
    B. Establish an IV with a crystalloid solution
    C. Transport patient to emergency department for antivenim
    D. Not apply ice

77. Many elderly people develop a "hunchback" appearance from rheumatoid arthritis, vertebral degeneration, and/ or poor posture. This exaggeration or angulation of the normal posterior curve of the thoracic spine is called:

    A. Londosis
    B. Gomphosis
    C. Kyphosis
    D. Ketosis

78. Which of the following medications is commonly used to treat patients who are of organoposphate poisoning?

    A. Atropine sulfate
    B. Flumanzenil
    C. Calcium chloride
    D. Adenosine

79. You are treating a 7 month old infant with a high fever. Upon arrival, the mother explains that the infant has had a recent upper respiratory and ear infection, your examination of the infant reveals that the infant is irritable, with fever, lethargy and a bulging anterior fontanelle. The father states that the infant has not been eating well at all. You would suspect:

    A. Bronchiolitis
    B. Meningitis
    C. Croup
    D. Epliglottis

80. What does the treatment for a patient whose EKG shows PVC's include?

    A. Immediate synchronized cardioversion with 50 – 100 joules
    B. Vagal maneuvers and 6 mg of adenosine rapid IV push over 1 – 3 seconds
    C. Observation only as long as the patient remains asymptomatic
    D. 1 – 1.5 mg/kg of lidocaine via slow IV push

81. You are treating an injured 4 year old child, the normal range of vital signs for that age group are most likely:

    A. Pulse 100 – 160 beats per minute, systolic blood pressure 50 – 70, respirations 30 – 60 breaths per minute
    B. Pulse 80 – 120 beats per minute, systolic blood pressure 80 – 110, respirations 20 – 30 breaths per minute
    C. Pulse 90 – 120 beats per minute, systolic blood pressure 80 – 100, respirations 25 – 40 breaths per minute
    D. Pulse 60 – 90 beats per minute, systolic blood pressure 90 – 120, respirations 15 – 20 breaths per minute

82. Type I diabetes diabetes mellitus is characterized by all of the following except:

   A. Metabolic acidosis
   B. Inadequate production of insulin
   C. Total body edema
   D. Accumulation of organic acids and ketons

83. Recommended pre-hospital medications for all patients with an AMI include _____ in the absence of contraindications.

   A. Beta-blockers
   B. Aspirin
   C. Nitroglycerin
   D. None of the above

84. When dispatched to a known violent scene, the paramedic should stage the vehicle:

   A. Directly behind a police vehicle
   B. At least 100 feet from the scene
   C. Out of sight of the scene
   D. At least 500 feet from the scene

85. Following a severe steering wheel chest impact, the pericardial space rapidly accumulates fluids resulting in pericardial tamponade. In the average size adult, tamponade can occur with as little as _____milliliters of fluid.

   A. 50 ml
   B. 100 ml
   C. 25 ml
   D. 75 ml

86. Which of the following factors causes a shift in the oxyhemoglobin curve, decreasing the tissue oxygen delivery?

   A. Increased metabolic rate
   B. Decreased body temperature
   C. Acidosis
   D. Increased body temperature

87. The force of the next cardiac contraction based on how much the cardiac muscle fiber stretchers is known as:

   A. Frank-Starling Law
   B. Cushing's reflex
   C. Beck's triad
   D. Kehr's sign

88. The kidneys are located in the:

   A. Pelvic cavity
   B. Thoracic cavity
   C. Retroperitoneal space
   D. Peritoneal space

89. The first step in triage at an MCI is to:

   A. Evaluate the victim's mental status and ABC's
   B. Direct the walking wounded away from the scene
   C. Assess the victim's hemodynamic status and AVPU
   D. Assess the victim's respiratory status and pulse rate

90. A patient begins to have a generalized seizure while running a
     marathon on a hot day. Which of the following procedures should
     you do first?

     A. Establish an airway and ventilate the patient
     B. Place cold packs around the neck and under the arms
     C. Move the patient into the ambulance
     D. Administer 5 mg of valium intravenously

91. The major extracellular cation is:

     A. Sodium Bicarbonate
     B. Calcium
     C. Sodium
     D. Potassium

92. The most common cause of upper airway obstruction is:

     A. Anaphylaxis
     B. Relaxation of the tongue
     C. Croup
     D. Epiglottitis

93. Your patient is a 38 year old woman who is eight months pregnant.
     You note that her blood pressure is 140/90 and edema is present
     all over her body. The patient is anxious and complains of seeing
     spots and having a headache. From this information, what condition
     should you suspect is present?

     A. Pre-eclampsia
     B. Gestational diabetes
     C. Hypertensive crisis
     D. Eclampsia

94. Endocrine glands do what?

    A. Produce adrenaline
    B. Excrete hormones into the blood stream
    C. Produce enzymes
    D. Stimulate the sympathetic system

95. In respiratory acidosis, CO_2 retention leads to increased levels of:

    A. SpO_2
    B. pCO_2
    C. O_2
    D. pO_2

96. All of the following are used to treat stress, except:

    A. Patient education
    B. Vaccination
    C. Psychotherapy
    D. Medication

97. Acquired hypersensitivity is called a/an:

    A. Cancer
    B. Allergy
    C. Rheumatic fever
    D. Asthma

98. During the shift from fetal circulation to normal infant circulation, which of the following occurs?

  A. The lungs develop
  B. The ductus venosus closes
  C. The ductus arteriosus opens
  D. The lungs develop expectorants

99. All of the following are included in the universal precautions except:

  A. Immediate vaccinations
  B. Hand washing
  C. Use of sharp containers to discard needles, syringes, etc.
  D. Wearing gloves and other barrier precautions

100. What does a carotid artery bruit indicate?

  A. Congestive heart failure
  B. Good peripheral perfusion
  C. Jugular vein distention
  D. Obstruction of blood flow

101. You would be likely to receive an order to administer intravenous thiamine to a patient who appeared to be:

  A. Hyperventilating
  B. In metabolic shock
  C. Profoundly intoxicated
  D. In status epilepticus

102.   How will the skin appear over a second degree burn?

    A.   Mottled red
    B.   Bright red
    C.   Charred black
    D.   Pearly white

103.   The acronym HEENT refers to:

    A.   Assessment of the peripheral extremities
    B.   Assessment of the pelvic girdle
    C.   Assessment of the head
    D.   Assessment of the belly

104.   When the APGAR mnemonic and scoring system is used to assess the newborn, the "P" stands for:

    A.   Pupils
    B.   Pulse
    C.   Pallor
    D.   Pink

105.   Which of the following behavioral disorders is generally the opposite of depression?

    A.   Suicide
    B.   Anxiety
    C.   Mania
    D.   Schizophrenia

106. What are the two basic goals of patient packaging?

    A. Stabilize and prepare for transport
    B. Thorough initial and rapid trauma assessments
    C. Secure the airway and stabilize the cervical spine
    D. Oxygenate and effective bleeding control

107. What is the correct method to stimulate respirations in a neonate?

    A. Hold it by the feet while you slap the buttocks
    B. Rub the head but avoid touching the fontanel
    C. Slap the soles of the feet and rub the back
    D. Let the cool air cause it to shiver a little

108. Early symptoms of overdose of tricyclic antidepressants include which of the following?

    A. Altered mental status and slurred speech
    B. Tachycardia and a wide QRS complex
    C. Nausea, vomiting, and severe diarrhea
    D. Psychosis and bizarre behavioral changes

109. Your patient is a 5 year old child who is conscious, but not breathing because of an airway obstruction. What is the first thing you should do for this patient?

    A. Open the airway with a head-tilt, chin lift
    B. Visualize the airway and perform a finger sweep
    C. Perform sub-diaphragmatic abdominal thrusts
    D. Give five back blows, followed by five chest thrusts

110. What does the term effacement refer to:

    A. Thinning of the cervix during the first stage of labor
    B. Opening of the cervix during the last stage of labor
    C. The position of the fetus in the uterus prior to birth
    D. The direction the fetus is facing during the birth

111. Information obtained from the patient and/or bystanders about the current event, including what led up to it is called:

    A. Positive feedback
    B. Focused history
    C. Chief complaint
    D. Signs and symptoms

112. The partial pressure of oxygen in the atmosphere at sea level is _____ torr.

    A. 0.3
    B. 597.0
    C. 159.0
    D. 3.7

113. In which are group are the kidneys unable to concentrate urine?

    A. Late adulthood
    B. Infancy
    C. Toddler
    D. Adolescence

114. Medical control has given you an order to administer 30 ml of syrup
      of ipecac to your patient who has ingested a poison. How many
      ounces do you administer to complete the order?

      A. One
      B. Four
      C. Two
      D. Three

115. The primary concerns in treating a near-drowning victim are
      management of:

      A. Aspiration and depressed brain function
      B. Hypoxia and acidosis
      C. Circulatory collapse and cardiac dysrhythmias
      D. Venous pooling and pulmonary embolus

116. Which of the following signs and symptoms requires immediate
      corrective action in the pre-hospital setting?

      A. Resting heart rate of 56 in an athlete
      B. Pulse of 106 in a child
      C. Decreased level of consciousness
      D. Night sweating in an AIDS patient

117. Which of the following best describes the pathophysiology of CHF?

      A. Cardiac muscle failure resulting in pulmonary edema
      B. Superior vena cava failure resulting in pulmonary edema
      C. Aortic valve failure resulting in pulmonary edema
      D. Pneumonia resulting in pulmonary edema

118. You have just started and IV lifeline, but the fluid is not flowing properly. What is the first thing you should do to troubleshoot this situation?

    A. Lower the IV bag below the level of the patient's arm
    B. Make sure the constricting band has been removed
    C. Ensure that the right size drip set is attached
    D. Remove the cannula and try another site

119. When intubating a child under the age of 8 years old, you should remember that:

    A. An uncuffed tube should be used
    B. A laryngoscope is not used
    C. A higher flow rate of oxygen is needed
    D. A bag valve device may not be used

120. All of the following drugs are used to treat pulmonary edema (CHF) except:

    A. Lasix
    B. Atropine
    C. Digitalis
    D. Nitroprusside

121. Which of the following is a disease that is associated with cigarette smoking and is related to, but distinct from, emphysema?

    A. Hemopneumothorax
    B. Congestive heart failure
    C. Chronic bronchitis
    D. Simple pneumothorax

122. Best sign to evaluate the neurological system is:

       A. Level of consciousness
       B. Loss of sensation
       C. Pupils
       D. Increased blood pressure

123. Your burn patient requires 1600 ml of normal saline over 8 hours. The drop factor is 10 gtts/min. How many gtts/min are required to infuse the IV at this rate?

       A. 33 gtts/min
       B. 25 gtts/min
       C. 65 gtts/min
       D. 44 gtts/min

124. Testosterone is a hormone secreted by the:

       A. Adrenal glands
       B. Testes
       C. Thyroid
       D. Ovaries

125. The symbol for chloride is:

       A. Cl+
       B. C-
       C. Cl-
       D. C+

ANSWER SHEET 2

1. C- The patient is hypovolemic
2. A- Phobia
3. C- Psychologic
4. A- Ingestion
5. A- 0.8 mg/ml
6. C- Anemia
7. C- Spinal injury
8. C- Thiamine
9. D- Epiglottitis
10. C- Calf
11. C- 102 degrees
12. A- 36 – 46
13. A- Trauma
14. A- Inadequate tissue perfusion
15. C- Suction the upper airway using a meconium aspirator
16. B- 50 ml's
17. B- Supine with head and neck hyperextended
18. D- During exercise or stress
19. B- Lowbrain
20. A- Dry lime
21. B- Repolarization of the ventriles
22. D- For each new patient
23. C- Shopping service
24. A- Monitor internal changes of the body
25. A- Bradycardia
26. A- Tremors and parasthesia
27. C- Change in mental status
28. B- Relieve cardiac compression
29. A- Separate the walking wounded from the most severely injured
30. A- Hydrogen ions in body fluids
31. C- Both are correct

32.   B- Trending mental status and vital signs
33.   A- Pulmonary edema
34.   A- T-10
35.   B- Explain to the patient the seriousness of her condition
36.   B- Darvon
37.   B- Fluid in the lungs
38.   C- Exercise or stress
39.   A- Hives
40.   C- Dyspnea, rales, and rhonchi
41.   A- Ingestion of fruits prior to obtaining a reading
42.   A- Kidneys
43.   C- Retinal artery occlusion
44.   A- The lower gastrointestinal tract
45.   A- A narcotic overdose
46.   C- Hematoma
47.   A- Epinephrine
48.   C- 7.5
49.   A- Megahertz
50.   A- 60 cc syringe

ANSWER SHEET 2

51. A- Think and work under pressure
52. B- Cyanosis
53. A- 6
54. D- Intubation may be avoided where it may have previously been required
55. A- Basic life support
56. D- First degree AV heart block
57. A- Immediate cardioversion
58. B- 15 seconds
59. A- 12 vertebrae
60. B- Transport the patient to the hospital
61. A- R wave
62. C- Immediately upon the birth of the baby
63. C- Nausea and vomiting
64. A- Hypothalamus
65. C- Apnea monitor
66. B- Rules, standards, and morals
67. B- Increased blood pressure and increased pulse
68. A- SVT
69. A- O-
70. A- Ischemia of the trachea wall
71. C- Stabilize the vehicle
72. A- 100 x 100
73. A- A high pitched sound upon inspiration from an airway obstruction
74. B- 4 million
75. C- 2 mg of morphine sulfate
76. A- Apply a constricting band\
77. C- Kyphosis
78. A- Atropine sulfate
79. B- Meningitis
80. C- Observation only as long as the patient remains asymptomatic

81. B- Pulse 80 – 120 beats per minute, systolic blood pressure 80 – 110, respirations 20 – 30 breaths per minute
82. C- Total body edema
83. B- Aspirin
84. C- Out of sight of the scene
85. A- 50 ml
86. B- Decreased body temperature
87. A- Frank-Starling Law
88. C- Retroperitoneal space
89. B- Direct the walking wounded away from the scene
90. A- Establish an airway and ventilate the patient
91. C- Sodium
92. B- Relaxation of the tongue
93. A- Pre-eclampsia
94. B- Excrete hormones into the blood stream
95. B- pCO2
96. A- Patient education
97. B- Allergy
98. B- The ductus venosus closes
99. A- Immediate vaccinations
100. D- Obstruction of blood flow

ANSWER SHEET 2

101. C- Profoundly intoxicated
102. A- Mottled red
103. C- Assessment of the head
104. B- Pulse
105. C- Mania
106. A- Stabilize and prepare for transport
107. C- Slap the soles of the feet and rub the back
108. B- Tachycardia and wide QRS complex
109. C- Perform sub-diaphragmatic abdominal thrusts
110. A- Thinning of the cervix during the first stage of labor
111. B- Focused history
112. C- 150.0
113. B- Infancy
114. A- One
115. C- Hypoxia and acidosis
116. C- Decreased level of consciousness
117. A- Cardiac muscle failure resulting in pulmonary edema
118. B – Make sure the constricting band has been removed
119. A- An uncuffed tube should be used
120. B- Atropine
121. C- Chronic bronchitis
122. A- Level of consciousness
123. A- 33 gtts/min
124. B- Testes
125. C- Cl-

PARAMEDIC STUDY QUESTIONS

Test 3

1. Common causes of laryngeal spasm include all of the following except:

   A. CVA
   B. An overly aggressive intubation attempt
   C. Post-extubation
   D. Cold water drowning

2. The radio report to the receiving hospital should include all of the following, except:

   A. The chief complaint
   B. The patient's name
   C. Baseline vital signs
   D. Response to emergency medical care

3. When confronted with an arm or leg presentation, the paramedic should:

   A. Transport the mother to the hospital immediately
   B. Gently manipulate the baby's body by the presenting part so that delivery is made easy.
   C. Pull the baby's body out by the presenting part
   D. Push the presenting part back into the vagina

4. Which infection is transmitted through contact with blood or body secretions?

    A. Varicella
    B. Hepatitis B
    C. Hepatitis A
    D. Tuberculosis

5. You are evaluating an EKG strip with no P waves and it is irregular. What is the most likely interpretation?

    A. Atrial flutter
    B. Ventricular tachycardia
    C. Atrial fibrillation
    D. 3rd degree heart block

6. Blood under the dura is described as:

    A. Subdural
    B. Contusion
    C. Epidural
    D. Subarachnoid

7. What is the purpose of performing the Sellick"s maneuver?

    A. To visualize the upper airway structures during BVM
    B. To prevent vomiting during attempts at intubation
    C. To prevent the tongue from blocking the airway
    D. To protect a patient with possible spinal injury

8. You are called to the scene of a 5 year old with sudden onset of shortness of breath. The child is dyspneic and has wheezing on the left chest area. What do you think is the cause of this child's distress?

    A. Asthma
    B. Pulmonary embolus
    C. Foreign body aspiration
    D. Pneumothorax

9. The senior citizen has special problems to her aging process: The correct description of a body change due to the aging process is:

    A. Increased thirst
    B. Enhanced pain response
    C. Diminished vision
    D. Enhancement of vision

10. You are assisting in a delivery in the field. As the baby's head is born, you realize that the umbilical cord is wrapped around the baby's neck. What is your first step in the management of this problem?

    A. Attempt to slip the cord over the baby's head
    B. Position the patient as for a prolapsed cord
    C. Apply two clamps and cut the cord immediately
    D. Moisten the cord and transport immediately

11. Secondary injuries from a MVC, where the driver went up and over the steering column, include:

    A. Cervical spine
    B. Face and head
    C. Chest and abdomen
    D. Head and neck

12. Muffled or distant heart sounds may indicate the presence of:

    A. Bruits
    B. Fluid
    C. Trapped air
    D. A broken stethoscope

13. Which of the following is not a component of the lower airway?

    A. Trachea
    B. Bronchi
    C. Pryiform fossae
    D. Carina

14. Any chemical substance that when taken into a living organism produces a biologic response affecting one or more of that organism's processes or functions is a :

    A. Drug
    B. Antibody
    C. Solution
    D. Antigen

15. Red blood cells are also known as:

    A. Erythrocytes
    B. Hemocytes
    C. Leukocytes
    D. Thrombocytes

16. All of the following drugs have analgesic, anti- inflammatory, and intipyretic properties, EXCEPT:

    A. Ketorolac
    B. Acetaminophen
    C. Aspirin
    D. Ibuprofen

17. A burn characterized by a moist, red appearance with blisters is probably _____.

    A. Fourth
    B. First
    C. Second
    D. Third

18. You are called to respond to a possible stroke patient. When you arrive at the patient's apartment, you find a 94 year old female who awoke feeling weak. On attempting to walk to the bedroom, the patient noted that her left leg was dragging, and her friend could not understand her on the telephone. On examining the patient, you note that her blood pressure is 186/120 bilaterally, her speech is truly slurred, and she has 2 out of 4 strength in her left arm and 1 out of 4 strength in her left leg All of the following are appropriate treatment options for this patient except:

    A. Request permission to open and administer a sublingual nifedipine capsule to lower the blood pressure.
    B. Establish and maintain the airway
    C. Administer oxygen and attach the patient to an EKG monitor
    D. Protect the paretic limbs in order to prevent injury during transport

19. Of the following, which is not a sign of pericardial tamponade?

  A. Rapid thready pulse
  B. Narrowing pulse pressure
  C. Tracheal deviation
  D. Jugular vein distention

20. A widened mediastinum on a chest x-ray is a hallmark finding of _____.

  A. Brucellosis
  B. Anthrax
  C. Plague
  D. Tularemia

21. The average weight of a newborn at birth is:

  A. 4.5 – 5.0 kg
  B. 3.0 – 3.5 kg
  C. 2.5 – 3.0 kg
  D. 4.0 – 4.5 kg

22. The average weight of an infant is double its birth weight by what age?

  A. One year
  B. 9 – 12 months
  C. 4 – 6 months
  D. 3 months

23. Which of the following devices would you use to inspect a patient's eyes?

    A. Ophthalmoscope
    B. Doppler
    C. Otoscope
    D. Hydroscope

24. Special radio codes:

    A. Add an unnecessary level of complexity
    B. Should be used in public
    C. Lead to a clearer understanding of the message
    D. Are necessary to avoid listeners

25. Cardiac muscle tissue has the ability to contract without neural stimulation. This property is called:

    A. Depolarization
    B. Automaticity
    C. Self-excitation
    D. Syncytin

26. In addition to administering Pitocin to a patient experiencing postpartum hemorrhage, the Paramedic should implement which of the following measures?

    A. Massage the fundus, IV fluids, and place baby on the breast
    B. IV fluids, pack the vagina and place mother in the shock position
    C. IV volume expanders, anti-shock trousers, and EKG monitor
    D. IV KVO. High-flow oxygen, and fundal massage

27. A normal property of the brain that maintains cerebral perfusion within a fairly wide range of mean arterial blood pressures is known as:

    A. Homeostasis
    B. Cerebral autoregulation
    C. Beck's triad
    D. Frank- Starling law

28. There are three major types of acute abdominal pain:

    A. Diffuse
    B. Referred
    C. Involuntary
    D. Voluntary

29. Which of the following components of kinetic energy make the greatest impact on a trauma patient?

    A. Velocity
    B. Deceleration
    C. Mass
    D. Acceleraton

30. The purpose of controlling a hemorrhage by direct pressure is to limit additional significant blood loss and to:

    A. Avoid the use of indirect pressure
    B. Limit exposure to communicable disease
    C. Promote localized clotting
    D. Stop the arterial blood flow

31. Which of the following do not commonly cause chemical burns?

    A. Saw dust
    B. Acids and bases
    C. Phenols
    D. Dry chemicals

32. Sensation in the inguinal crease is located in which nerve root?

    A. S-1
    B. C-8
    C. L-2
    D. T-4

33. Which of the following is the immediate concern for the paramedic when treating a child that has a poisoning or drug overdose?

    A. Respiratory depression
    B. Shock
    C. Anaphylaxis
    D. Vomiting and aspiration

34. Which of the following GI tract devices is not commonly in the home care setting?

    A. Colostomy bags
    B. P-tubes
    C. J-tubes
    D. G-tubes

35. The rescue term that means to remove the debris or parts of the vehicle from around the patient, so the patient may be freed for removal is:

    A. Gaining access
    B. Evacuation
    C. Disentanglement
    D. Extrication

36. Which of the following tools might be used to open a vehicle door jammed after a collision?

    A. Hydraulic spreaders
    B. Halogen tool
    C. Come-alongs
    D. Pry bar

37. The pediatric dose of epinephrine 1:10,000 is 0.1 ml/kg which yields:

    A. 10.0 mg/kg
    B. 0.1 mg/kg
    C. 0.01 mg/kg
    D. 1.0 mg/kg

38. The approximate percentage of oxygen found in room air is:

    A. 28%
    B. 19%
    C. 21%
    D. 28%

39. The primary use of the Magill forceps in the field is to:

     A. Aid in removal of an esophageal obturator airway in the field
     B. Directly remove a visible foreign body obstruction
     C. Open the airway of a patient with suspected neck trauma
     D. Move the tongue aside during attempts at intubation

40. When responding to the scene of a hazardous material incident, the EMS crew should approach from:

     A. Uphill and upwind
     B. Downhill and downwind
     C. Downhill and upwind
     D. Uphill and downwind

41. The pain caused by myocardial infarction is usually relieved only by the use of which of the following medications?

     A. Oxygen
     B. Morphine
     C. Nitroglycerin
     D. Acetaminophen

42. Which heart sounds are normal findings on auscultation?

     A. S2 and S4
     B. S1 and S2
     C. S2 and S3
     D. S3 and S4

43.   Insufficient oxygenation:

   A.   Increases the respiratory rate
   B.   Decreases the respiratory rate
   C.   Has no effect on the respiratory rate
   D.   Any of the above

44.   Anything that breaks homeostasis is defined as:

   A.   Insult
   B.   Infarction
   C.   Stress
   D.   Injury

45.   Pathophysiology is the:

   A.   Study of cells
   B.   Study of structure
   C.   Study of disease
   D.   Study of function

46.   The straight blade for the laryngoscope is called the:

   A.   Miller
   B.   Macintosh
   C.   McCoy
   D.   Morgan

47.   Paramedics who respond to back country rescues should be trained in all of the following, except:

    A.   Hyperbaric therapy
    B.   Long-term hydration management
    C.   Reposition dislocations
    D.   Manage hyperthermia

48.   If you have to retreat from a dangerous scene, be sure to:

    A.   Bring cover with you
    B.   Make sure the dispatcher does not send any further EMS units to the scene
    C.   Bring the patient with you
    D.   Document that you did not abandon the patient

49.   You are caring for a patient who is very angry and is taking her anger out on you. Which of the following should you avoid doing?

    A.   Request police
    B.   Do a quick pat down for weapons
    C.   Getting angry in return
    D.   Keep the situation calm

50.   The pediatric dose of Lidocaine is:

    A.   1 mg/kg of body weight
    B.   2 mEq/kg of body weight
    C.   1 mEq/kg of body weight
    D.   0.5 mg/kg of body weight

51. Care for the patient with a cardiac contusion is similar to care for the patient with which or the following conditions?

    A. Myocardial infarction
    B. Closed abdominal trauma
    C. Pericardial tamponade
    D. Tension pneumothorax

52. Which of the following is not an example of an abnormal breathing pattern?

    A. Agonal
    B. Eupnea
    C. Biot's
    D. Central neurogenic hyperventilation

53. Suctioning can cause vagal stimulation by:

    A. Irritating the carotid sinuses
    B. Changing the intrathoracic pressure
    C. Tickling the back of the throat
    D. Creating a hypoxic state

54. The acronym "AEIOU-TIPS" is helpful to the paramedic when considering the causes of:

    A. Altered mental status
    B. Shortness of breath
    C. Chest pain
    D. Abdominal pain

55. This airway is measured from the corner of the mouth, to the tip of the ear lobe on the same side:

    A. Nasopharyngeal
    B. Endotracheal
    C. Oropharyngeal
    D. Nasotracheal

56. Medical control orders 5 mg of Valium to be administered to a patient. The Valium ampule contains 10 mg of Valium in 2 ml of solvent. How many ml's would you administer?

    A. 1.0 ml
    B. .001 ml
    C. .0` ml
    D. 10 ml

57. Before testing an unconscious head-injury patient's response to pain, the paramedic should:

    A. Test for grip strength
    B. Protect the neck
    C. Examine the abdomen
    D. Test the lower extremity reflexes

58. In tachycardia, the rate is:

    A. 80 – 120
    B. Below 60
    C. Above 100
    D. 40 – 60

59. The adjunctive procedure which is most successful in resuscitating the cardiac patient it:

    A. Defibrillation
    B. A mechanical chest compressor
    C. The bag-valve mask
    D. The demand valve

60. Venous access not usually available in the adult patient, but used for the pediatric one is the:

    A. Intraosseous site
    B. Jugular vein
    C. Femoral vein
    D. Scalp vein

61. Cells that secrete collagen contributing to wound healing are called:

    A. Astrocytes
    B. Osteocytes
    C. Fibroblasts
    D. Fibroclasts

62. Laryngeal Mask Airways (LMA) is a pear shaped device placed into the:

    A. Larynx
    B. Hypopharynx
    C. Esophagus
    D. Trachea

63. Cellular division is also known as:

     A. Mitosis
     B. Symbiosis
     C. Osmosis
     D. Diffusion

64. Which drug is used in management of congestive heart failure?

     A. Verapamil
     B. Bretylium
     C. Dobutamine
     D. Isoproterenol

65. Until ruled out by a physician, documented fever in an infant younger
     than three months old is always considered to result from which of
     the following?

     A. Meningitis
     B. Epilepsy
     C. Reye's syndrome
     D. Epiglottitis

66. All of the following are acute emergencies that may progress to
     cardiac arrest, except:

     A. Gunshot wound
     B. Acute MI
     C. Viral flu syndrome
     D. Foreign body airway obstruction

67. The documentation format that follows patient care in the order it was accomplished is known as the:

    A. Data entry format
    B. Chronological format
    C. Accepted format
    D. Consumption format

68. Which type of hepatitis is spread via the oral or fecal route?

    A. Hepatitis D
    B. Hepatitis A
    C. Hepatitis C
    D. Hepatitis B

69. Which of the following drugs is administered only by inhalation?

    A. Epinephrine 1:10,000
    B. Terbutaline sulfate
    C. Albuterol
    D. Aminophylline

70. What is the purpose of offline medical protocols?

    A. To provide a standardized approach to common patient problems
    B. To allow the paramedic to train less experienced EMS providers
    C. To give ultimate responsibility to the medical control physician
    D. To remove the possibility of malpractice from the EMS provider

71. You are dispatched to a 76 year old male with acute abdominal pain. As you arrive at the patient's home, the patient admits to having a long standing history of arthritis, which he self-medicates with Aspirin and Advil. He notes that the arthritis has been particularly painful for the past 5 weeks, and he states that he has increased the use of both of these medications. The patient notes that his pain has been in the epigastric area for the past 4 days and that it is a burning associated with vomiting and belching. Your physical assessment reveals vital signs as follows: Blood pressure- 149/86, Pulse rate- 112, Respirations- 22 breaths per minute. The patient is pale and diaphoretic and appears to be in mild distress. The abdominal examination reveals a non-distended abdomen, active bowel sounds, and epigastric tenderness without guarding or rebound. This presentation best represents which of the following diagnoses?

     A. Acute gastritis
     B. Acute appendicitis
     C. Acute kidney stone
     D. Acute gastrointestinal bleeding

72. In some cases of _____, priapism is an indicator of injury.

     A. Cardiac tamponade
     B. Airway obstruction
     C. Spinal injury
     D. Hypovolemia

73. The agency that develops rules and regulations for the use of all radio equipment is:

    A. The individual EMS agency
    B. Federal Communications Commission (FCC)
    C. Department of health
    D. Department of transportation

74. An oropharyngeal airway may be used to maintain an airway in a conscious patient:

    A. False
    B. True

75. When assessing a patient with dyspnea, you can be confident that he is adequately oxygenated if he does not have cyanosis.

    A. False
    B. True

76. The vocal cords are located in the:

    A. Pharynx
    B. Bronchi
    C. Larynx
    D. Carina

77. You are assigned to an unconscious diabetic patient. After your assessment, you administer oxygen and 50% dextrose intravenously (IV). The patient awakens and is transported to the nearest hospital without incident. You initially treated this patient under which type of consent?

    A. Implied
    B. Informed
    C. Assumed
    D. Involuntary

78. You are dispatched to the scene where a 55 year old male pedestrian has been hit by a car. In addition to your ambulance, a fire department engine company was also dispatched with another level of EMS professional to assist in provision of treatment. Which of the following best defines this person's role?

    A. First Responder
    B. Family member
    C. School nurse
    D. EMT- Intermediate

79. A group of cells that perform in a similar function are called:

    A. Diffusion
    B. Mitosis
    C. Symbiosis
    D. Osmosis

80. The gland that prepares the body to react rapidly to acute stress is the:

    A. Thyroid
    B. Pituitary
    C. Adrenal medulla
    D. Parathyroid

81. Paramedics should always _____ before they perform a procedure on a patient:

    A. Speak to the patients family
    B. Explain what they are going to do
    C. Discuss another topic to distract the patient
    D. Contact medical control

82. For children, the Paramedic should apply compressions with:

    A. The heel of one hand
    B. The finger tips
    C. Both hands
    D. One finger

83. Which of the following is most suggestive of a hemorrhagic gastrointestinal problem:

    A. Diarrhea
    B. Melena
    C. Diaphoresis
    D. Chest pain

84. Occasionally a drug is dose dependent on the weight of the patient. If you were to have a patient weighing 220 pounds, how many kilograms would this patient weigh?

    A. 22 kg
    B. 100 kg
    C. 1000 kg
    D. 220 kg

85. When the body forms a "memory" for certain antigens, this is referred to as/ an:

    A. Acquired immunity
    B. Native immunity
    C. Anamnestic response
    D. Acquired immunity

86. Ethics are defined as:

    A. A system of principles governing moral conduct
    B. A true measure of honesty
    C. Not acting in a rude or crude manner
    D. Never cheating on your taxes

87. Decreased perfusion can be caused by an event resulting in blood loss, kinking of the great vessels or:

    A. By release of antidiuretic hormone
    B. Loss of vasomotor tone
    C. An allergic reaction
    D. A failure in the buffer system

88. Traumatic injuries associated with consensual sex include all of the following, except:

    A. Restraint injuries
    B. Fractured penis
    C. Emotional withdrawal
    D. Tears from pierced body jewelry

89. The most common causes of respiratory distress in the infant age group include all the following, except:

    A. Epiglottitis
    B. Asthma
    C. Croup
    D. Bronchiolitis

90. A QRS complex is considered abnormal if it lasts longer that how many seconds?

    A. 0.04 seconds
    B. 0.12 seconds
    C. 0.08 seconds
    D. 0.10 seconds

91. Traumatic aortic rupture usually occurs as a result of transaction of the aorta at the:

    A. Pulmonary artery
    B. Ligamentum arteriosum
    C. Ligamentum teres
    D. Ligamentum nuchae

92. A patient presents with symptoms of flushing, itching, hives, difficulty breathing, decreased blood pressure, and dizziness. What should you suspect?

    A. Anaphylaxis
    B. Stroke
    C. Acute appendicitis
    D. Diabetic coma

93. The most dangerous complication of a neck vein laceration is:

    A. Ecchymosis of the neck
    B. A blood clot
    C. An air embolus
    D. Carotid artery compression

94. The amount of air in each breath is called:

    A. Alveolar air
    B. Tidal volume
    C. Stroke volume
    D. Dead air

95. An adult who has burns over both sides of one arm and both sides of one leg would be estimated to have total burns over what percentage of body surface area?

    A. 27%
    B. 36%
    C. 9%
    D. 18%

96. Which of the following is not a sign of pure right sided heart failure?

  A. Pedal edema
  B. Hepatomegaly
  C. Pulmonary edema
  D. Positive hepatojugular reflex

97. Ventilating a patient at 30 breaths per minute with a bag-valve and high-flow oxygen may result in:

  A. Alkalizing the bloodstream
  B. Production of a tension pneumothorax
  C. An increase in intracranial pressure
  D. A decrease in the normal serum blood pH

98. With whom does ultimate responsibility for patient care in the field always rest?

  A. Whomever provides online direction
  B. The medical control physician
  C. The highest trained provider on scene
  D. The regional or state EMS director

99. You are transporting a 28 year old multigravida patient whose last two babies were delivered by Cesarean section. She appears to be near full term and is having hard contractions 3 – 4 minutes apart. She suddenly complains of continuous lower abdominal pain, progressively worse, and gradually goes into shock. The most likely possibility is:

  A. Uterine rupture
  B. Placenta previa
  C. Abruptio placents
  D. Ruptured ectopic pregnancy

100.   Which of the following drugs is an antidysrhythmic agent?

   A.   Furosemide
   B.   Isoproterenol
   C.   Lidocaine
   D.   Nitroglycerin

101.   The number one cause of death in children is:

   A.   SIDS
   B.   Drowning
   C.   Trauma
   D.   Suffocation

102.   The most dangerous disorder causing pediatric upper airway stridor
        would be:

   A.   Status asthmaticus
   B.   Croup
   C.   Epiglottitis
   D.   Peritonsilar abscess

103.   During the physical examination the older patient must be handled
        gently so as not to:

   A.   Force the patient to receive unwanted care
   B.   Cause any additional injury
   C.   Overwhelm the caretaker with your techniques
   D.   Confuse the patient with your specialized equipment

104. In the year 2030, it is projected that one in _____ people will be age 65 or older in the United States.

    A. Five
    B. Twenty- five
    C. Three
    D. Ten

105. Hospice care emphasizes comfort measures and counseling to provide physical, spiritual, social, and _____ needs.

    A. Financial
    B. Long distance
    C. Economic
    D. Long-term

106. A very useful aspect to the use of triage tags at an MCI is that:

    A. They are inexpensive
    B. They help to eliminate the need to reassess each patient over and over again
    C. They do not take any training to use
    D. Patients can fill them out themselves

107. The principles of _____ medicine classifies interventions into one of five categories:

    A. Osteopathic
    B. Diagnostic
    C. Evidence based
    D. Homeopathic

108.   Diabetes Ketoacidosis is best treated by:

    A. A continuous insulin infusion
    B. Administering oral hypoglycemics
    C. A wide open IV of D5W
    D. A prophylactic Normal saline drip

109.   Complaints of abdominal pain are associated with:

    A. Hypoglycemia
    B. Hyperglycemia or diabetic ketoacidosis
    C. Hypoglycemia or diabetic ketoacidosis
    D. Hyperglycemia

110.   The most important functional unit of the respiratory system is the:

    A. Bronchi
    B. Trachea
    C. Alveoli
    D. Epiglottis

111.   When inserting an esophageal obturator airway device, the head should be in which of the following position's?

    A. Extension
    B. Sniffing
    C. Neutral
    D. Flexion

112. In the pre-hospital setting, all of the following are parts of the abdominal examination of the patient complaining of abdominal pain except:

    A. Auscultation for bowel sounds
    B. Inspection
    C. Palpation
    D. Determination of vital signs

113. In treating a patient with a cocaine overdose, all of the following medications may be given, if indicated, except:

    A. IV lidocaine for ventricular tachycardia with a pulse
    B. IV flumazenil (Mazecon) for a concurrent diazepam (Valium) overdose
    C. Nitroglycerin for chest pain
    D. Benzodiazepines, such as midazolam ((Versed) or lorazepam (Ativan) overdose

114. You respond to the scene of an explosion. You are treating a 27 year old male who was approximately 30 feet from the explosion of a small backpack. Your assessment reveals that he has dozens of small round holes all over his body. You anticipate that this was caused by:

    A. Shrapnel
    B. Flying glass
    C. Pressure wave
    D. Tattooing from the explosive device

115. Which of the following factors would normally cause a decrease in a patient's respiratory rate?

  A. Hypoxia
  B. Sleep
  C. Anxiety
  D. Fever

116. An injury to the brain opposite the site of a blunt force impact is called:

  A. Coup force
  B. Contralateral
  C. Contrecoup
  D. Concussion

117. A patient with a suspected MI but without respiratory compromise should receive oxygen by what rate and delivery device?

  A. 3 – 4 L/min by nasal cannula
  B. 15 – 20 L/min by bag-valve mask
  C. 5 – 10 L/min by simple face mask
  D. 10 – 15 L/min by non-rebreather mask

118. Neonates are defined as:

  A. A preterm newborn child
  B. An infant up to 1 month old
  C. A term newborn from time of birth to 1 week old
  D. A infant up to 3 months old

119. All of the following are known to be major risk factors predisposing to coronary artery disease except:

    A. Diabetes
    B. Hypertension
    C. Marijuana use
    D. Older age

120. A soft-tissue injury that results when a joint is moved beyond its normal range of motion is known as a:

    A. Contusion
    B. Sprain
    C. Strain
    D. Dislocation

121. The vocal cords are located in the:

    A. Larynx
    B. Pharynx
    C. Bronchi
    D. Carina

122. Bronchioles are composed of:

    A. Septal cells
    B. Cartilaginous rings
    C. Smooth muscle
    D. Alveolar ducts

123. The cheek bones are called:

    A. Zygoma
    B. Spheniods
    C. Orbits
    D. Occiput

124. A drug used to kill a bacteria's growth or to decrease it is called:

    A. Antibiotic
    B. Anti-inflammatory
    C. Anti-anginal
    D. Antibacterial

125. Death secondary to myocardial infarction is MOST COMMONLY due to:

    A. Cardiac tamponade
    B. Pulmonary embolism
    C. Dysrhythmias
    D. Low blood pressure

ANSWER SHEET 3

1.   A- CVA
2.   B- The patient's name
3.   A- Transport the mother to the hospital immediately
4.   B- Hepatitis B
5.   C- Atrial fibrillation
6.   A- Subdural
7.   B- To prevent vomiting during attempt at intubation
8.   A- Asthma
9.   C- Diminished vision
10. A- Attempt to slip the cord over the baby's head
11. C- Chest and abdomen
12. B- Fluid
13. C- Pryiform fossae
14. A- Drug
15. A- Erythrocytes
16. B- Acetaminophen
17. C- Second
18. A- Request permission to open and administer a sublingual nifedipine capsule to lower the blood pressure
19. C- Tracheal deviation
20. B- Anthrax
21. B- 3.0 – 3.5 kg
22. C- 4 – 6 months
23. A- Ophthalmoscope
24. A- Add an unnecessary level of complexity
25. B- Automaticity
26. A- Massage the fundus, IV fluids, and place baby on the breast
27. B- Cerebral autoregulation
28. B- Referred
29. A- Velocity
30. C- Promote localized clotting

31.   A- Saw dust
32.   C- L-2
33.   A- Respiratory depression
34.   B- P-tubes
35.   C- Disentanglement
36.   A- Hydraulic spreaders
37.   B- 0.1 mg/kg
38.   C- 21%
39.   B- Directly remove a visible foreign body obstruction
40.   A- Uphill and upwind
41.   B- Morphine
42.   B- S1 – S2
43.   A- Increases the respiratory rate
44.   C- Stress
45.   C- Study of disease
46.   A- Miller
47.   C- Reposition dislocations
48.   B- Make sure the dispatcher does not send any further EMS unit to the scene
49.   C- Getting angry in return
50.   A- 1mg/kg of body weight

ANSWER SHEET 3

51. A- Myocardial infarction
52. B- Eupnea
53. C- Tickling the back of the throat
54. A- Altered mental status
55. C- Oropharyngeal
56. A- 1.0 ml
57. B- Protect the neck
58. C- Above 100
59. A- Defibrillation
60. A- Intraosseous site
61. C- fibroblasts
62. B- Hypohparynx
63. A- Mitosis
64. C- Dobutamine
65. A- Meningitis
66. C- Viral flu syndrome
67. B- Chronological format
68. B- Hepatitis A
69. C- Albuterol
70. A- To provide a standardized approach to common patient problems
71. A- Acute gastritis
72. C- Spinal injury
73. B- Federal Communication Commission (FCC)
74. A- False
75. A- False
76. C- Larynx
77. A- Implied
78. A- First responder
79. B- Mitosis
80. C- Adrenal medulla
81. B- Explain what they are going to do

82. A- The heel of one hand
83. B- Melena
84. B- 100 kg
85. C- Anamnestic response
86. A- A system of principles governing moral conduct
87. B- Loss of vasomotor tone
88. C- Emotional withdrawal
89. A- Epiglottitis
90. B- 0.12 seconds
91. C- Ligamentum teres
92. A- Anaphylaxis
93. B- A blood clot
94. B- Tidal volume
95. A- 27%
96. C- Pulmonary edema
97. A- Alkalizing the blood stream
98. B- The medical control physician
99. A- Uterine rupture
100. C- Lidocaine

ANSWER SHEET 3

101. C- Trauma
102. B- Croup
103. B- Cause any additional injury
104. A- Five
105. C- Economic
106. B- The help to eliminate the need to reassess each patient over and over again
107. C- Evidence based
108. A- A continuous insulin infusion
109. B- Hyperglycemia or diabetic ketoacidosis
110. C- Alveoli
111. C- Neutral
112. A- Auscultation for bowel sounds
113. B- IV flumazenil (Mazecon) for a concurrent diazepam (Valium) overdose
114. A- Shrapnel
115. B- Sleep
116. C- Contrecoup
117. A- 3 – 4 L/min by nasal cannula
118. B- An infant up to 1 month old
119. C- Marijuana use
120. B- Sprain
121. A- Larynx
122. C- Smooth muscle
123. A- Zygoma
124. A- Antibotic
125. C- Dysrhythmias

Test 4

1. List in order, the structures of the heart as blood would pass beginning at the right atrium: 1. Left ventricle, 2. Pulmonary valve, 3. Pulmonary vein, 4. Pulmonary artery, 5. Tricuspid valve.

    A. 4,3,2,5,1
    B. 5,2,4,3,1
    C. 1,3,4,2,5
    D. 2,4,3,5,1,

2. A behavioral emergency only occurs when a person is:

    A. Unable to cope
    B. Insane
    C. Neurotic
    D. Experiencing a loss

3. Which of the following medications is not administered by a paramedic to a patient in anaphylaxis?

    A. Vasopressor
    B. Bronchodilator
    C. Histamine
    D. Steroid

4. The phase of a seizure characterized by a loss of consciousness with muscle contraction is called:

    A. Clonic
    B. Tonic
    C. Preictal
    D. Aura

5. Disadvantages of the use of a gastric tube include all of the following, except:

    A. It interferes with a mask seal
    B. It may cause the patient to vomit
    C. The patient can still talk
    D. It may cause bradycardia

6. Oxygen cylinders need to be hydrostatically tested every five years. However, if the tank has a star after the test date, it is good for a (an) _____ year period.

    A. 10
    B. 8
    C. 15
    D. 12

7. During an inflammation response the substance released by mast cells is:

    A. Lymphocytes
    B. Steroids
    C. Histamine
    D. Cortisol

8. A paramedic found to be purposely hurting patients by withholding essential medications would be charged in a (an) court:

    A. Criminal
    B. Administrative
    C. Small claims
    D. Civil

9. Which of the following is least important in the care of a newborn:

    A. Maintain body temperature
    B. Monitor EKG
    C. Monitor respirations
    D. Clear airway

10. Materials safety data sheets contain which of the following?

    A. Local reporting telephone numbers
    B. Evacuation radius
    C. Melting and boiling point
    D. Ingestion antidotes

11. A related disease or condition that can result in rapid and total airway obstruction is:

    A. Laryngitis
    B. Epiglottitis
    C. Bronchitis
    D. Bronchiolitis

12. Education systems in EMS are striving to develop which of the following?

     A. National core contents to replace EMS program cirricula
     B. Bridging and transitioning EMS programs with all health profession's education
     C. Accreditation for EMS education programs
     D. All of the above

13. The key strategy for reducing deaths from motor vehicles is:

     A. Getting everyone in the vehicle to wear seat belts
     B. Reducing speed limits in school zones
     C. Advocating for more vehicles to include airbags
     D. Mandating lower speed limits on the highways

14. Change 44% into a decimal.

     A. 44.0
     B. 0.44
     C. 0.04
     D. 04.4

15. Basic life support is often used interchangeably with the terms (s).

     A. CPR
     B. BLS
     C. BCLS
     D. Any of the above are correct

16. The elderly often have an ineffective cough reflex caused by:

    A. Atrophy of the pituitary gland
    B. Exposure to cigarette smoke
    C. Lower production of cortisor
    D. Weakening of the chest wall

17. Which of the following statements about retrograde intubation is not correct?

    A. It should only be performed by properly trained providers
    B. It involves placing a needle through the cricothyroid membrane
    C. It involves the use of a guide wire
    D. It is a surgical procedure

18. Communication techniques that may be useful during the patient interview include all the following, except:

    A. Explanation
    B. Encoding
    C. Listening
    D. Silence

19. All of the following are examples of stable patients, except:

    A. Minor isolated injury
    B. Low grade fever
    C. Minor illness
    D. Significant MOI with neck injury

20. Benefits of protocols standing orders, and patient care algorithms include all of the following, except they:

    A. Provide some legally defensible backup for the paramedic
    B. Cover all aspects of the critically ill patient
    C. Promote a standard approach to patient care
    D. Clearly define performance parameters

21. The primary motor cortex located in the _____ is connected with the association motor cortex in the basal ganglia and cerebellum.

    A. Parietal lobe
    B. Occipital lobe
    C. Midbrain
    D. Frontal lobe

22. Appendicitis is inflammation of the appendix caused by occlusion of the lumen by a:

    A. Large lesion
    B. Small piece of stool
    C. Small tumor
    D. Large lesion

23. Which of the following is the dominant pacemaker of the heart?

    A. AV junction
    B. AV node
    C. Purkinje network
    D. SA node

24. Isoproterenol will:

     A. Increase the myocardial oxygen consumption
     B. Decrease the heart rate
     C. Decrease the cardiac output
     D. Increase AV conduction, but not effect ventricular
         conduction

25. All patient's in respiratory distress should receive oxygen:

     A. True
     B. False

26. Which of the following is the most important factor in determination
     of the mechanism of injury?

     A. The vehicle collides with another object
     B. The organs collide with the interior of the occupants
     C. The interior of the vehicle collides with the exterior
     D. The occupants collide with the interior of the vehicle

27. Claudication is defined as:

     A. Focal headache
     B. Abdominal cramps
     C. Intermittent heat and redness of a leg
     D. Cramp-like pain in the calf

28. What are the classic symptoms of narcotic overdose?

     A. Altered mental status, euphoria, and dilated pupils
     B. Respiratory depression and constricted pupils
     C. Cardiac dysrhythmias and altered mental statues
     D. Excitability, hyperactivity, and hypertension

29. Which of the following drugs does NOT commonly cause toxicity in elderly patients?

     A. Lidocaine
     B. Theophylline
     C. Digitalis
     D. Nitroglycerin

30. What is the most common complication of an acute myocardial infarction?

     A. The onset of cardiogenic shock
     B. Increased intermittent chest pain
     C. Occurrence of unstable angina
     D. The onset of a dysrhythmia

31. At what age is a person considered capable of giving consent to treatment?

     A. 21
     B. 15
     C. 17
     D. 18

32. A cardiac contraction rate of 60 to 100 times per minute is most commonly associated with which of the following locations?

   A. SA node
   B. Ventricles
   C. Atria
   D. AV junction

33. A 49 year old female is complaining of chest pain and shows premature ventricular complexes on the monitor. Which of the following medical management procedures is MOST indicated for this patient?

   A. Procainamide
   B. Oxygen
   C. Furosemide
   D. Morphine

34. After sealing an open chest wound, you notice that your patient is developing progressive dyspnea. You should do which of the following?

   A. Bag-valve mask the patient
   B. Administer a higher concentration of oxygen
   C. Release one corner of the occlusive dressing
   D. Rush the patient immediately to the hospital

35. Tetanus may be a complication of any open soft tissue injury and is caused by a:

   A. Disease
   B. Parasite
   C. Virus
   D. Soil bacterium

36.   The delivery of the placenta signifies:

      A.   The end of the third stage of labor
      B.   The beginning of the fourth stage of labor
      C.   The end of the second stage of labor
      D.   The beginning of the first stage of labor

37.   What drug can be given as an adjunct to Epinephrine in anaphylactic shock?

      A.   Calcium
      B.   Diphenhydramine
      C.   Valium
      D.   Calan

38.   Dosage for Benadryl is:

      A.   10 to 50 mg IV
      B.   25 to 50 mg IM
      C.   10 to 50 mg IV
      D.   13 to 60 mg IM

39.   The receptor sites that need to be stimulated to promote bronchodilation are:

      A.   Dopaminergic
      B.   Beta 2
      C.   Alpha
      D.   Both B and C

40. The onset of action of a drug is:

     A. A concentration that causes toxic levels
     B. The minimum effective concentration
     C. The maximum dose that can be given
     D. The difference between therapeutic index and the toxic
         levels

41. An acquired resistance to the therapeutic effects of usual doses of a
     drug is referred to as:

     A. Addiction
     B. Drug abuse
     C. Dependence
     D. Tolerance

42. If your patient was exposed to a herbicide, which antidote may prove
     useful if authorized by Medical Control to administer?

     A. Nitrous oxide
     B. Atropine
     C. Narcan
     D. Solu-medrol

43. Conditions that may contribute to hypothermia include all of the
     following, except:

     A. Hypothyroidism
     B. Brain Dysfunction
     C. Hypoglycemia
     D. Hyperglycemia

44. Which of the following are the actual units for kinetic energy?

   A. Foot- pounds
   B. Inch- ounces
   C. Mile- grams
   D. Meter- liter

45. One tablespoon is equivalent to:

   A. 30 cc
   B. 3 cc
   C. 15 cc
   D. 7.5 cc

46. Suicide attempts are most successful for:

   A. Persons under the age of 15
   B. Persons over the age of 45
   C. Females
   D. Males

47. Anaerobic metabolism can cause

   A. Metabolic acidosis
   B. Metabolic alkalosis
   C. Respiratory acidosis
   D. Respiratory alkalosis

48.   What is the usual pediatric dose of naloxone for a child less than five years old?

   A. 0.01 mg/kg via IV bolus
   B. 2 mg via IV bolus
   C. 0.1 mg/kg via IV bolus
   D. 1 mg via IV bolus

49.   What is the prehospital treatment for patients with suspected deep venous thrombosis?

   A. Immobilization and elevation of the extremity
   B. Treatment for shock, including oxygen and PASG
   C. Initiation of nitroglycerin and heparin
   D. Initiation of thrombolytic therapy to control the clot

50.   Oral administration of a "sports" drink or other beverage containing sodium is appropriate to the treatment of:

   A. Heat exhaustion
   B. Heat stroke
   C. Heat cramps
   D. Both A and C

51.   Meningitis is an infectious nervous system disease that is caused by:

   A. A virus
   B. Bateria
   C. Fungi
   D. All of the above

52. If your patient has an open abdominal wound with a loop of bowel obtruding, you should treat this with:

   A. A clean gauze dressing secured with sterile tape
   B. A trauma dressing secured with triangular bandages
   C. A wet sterile dressing and an occlusive dressing
   D. An occlusive dressing secured on only three sides

53. Depression is an example of a (an):

   A. Organic disease
   B. Psychiatric illness
   C. Psychosis
   D. Mood disorder

54. Paramedics should always _____ before they perform a procedure on a patient.

   A. Explain what they are going to do
   B. Speak to the patients family
   C. Discuss another topic to distract the patient
   D. Contact medical control

55. In dealing with the disturbed patient, the EMS provider should NEVER:

   A. Confront the patient
   B. Lie to the patient
   C. Restrain the patient
   D. Agree with the patient

56. In the United States about _____ people die fro electrical shock each year:

    A. 100,000
    B. 100
    C. 1,000
    D. 10,000

57. Which of the following corticosteroids is not used in anaphylaxis by the paramedic to slow histamine release and capillary leakage?

    A. Nortriptyline
    B. Aminophylline
    C. Solu-Medrol
    D. Hydrocortisone

58. The pacemaker cells in the _____of the heart normally initiate the electrical impulses that start the sequence of excitation and conduction through the heart.

    A. Purkinje fibers
    B. Bundle branches
    C. SA node
    D. AV node

59. The use of the AED is encouraged for all patients over the age of:

    A. 4
    B. 18
    C. 8
    D. 25

60. Drugs that have high potential for abuse and a extremely high dependence liability, but also have accepted medical uses are:

    A. Schedule II drugs
    B. Schedule I drugs
    C. Schedule IV drugs
    D. Schedule III

61. The contraction phase of the cardiac cycle is called:

    A. Tocus
    B. Milieu
    C. Diastole
    D. Systole

62. The greatest amount of intravascular fluid loss occurs with the first _____after the burn:

    A. 3 hours
    B. 1 hour
    C. 10 minutes
    D. it depends on the severity of the burn

63. Paradoxical chest wall movement is a sign of:

    A. AMI
    B. Guarding
    C. Two broken ribs
    D. Pneumothorax

64. What communication system has the capability to send and receive voice and telemetry simultaneously?

    A. Multiplex
    B. Duplex
    C. VHF
    D. UHF

65. A patient with left shoulder pain may have a:

    A. Ruptured spleen
    B. Bowel obstruction
    C. Pneumothorax
    D. Pelvic fracture

66. A patient has fallen off a 25 foot ladder, striking his back on a railing. He is experiencing pain at the injury site and a loss of bladder control. Which part of the spinal cord is most likely affected by this mechanism?

    A. Sacral
    B. Thoracic
    C. Cervical
    D. Lumbar

67. You are the first paramedic unit to arrive on the scene of a multi-injury bus accident. What is your first responsibility?

    A. Review and evaluate the efficiency of site operations up until your arrival
    B. Extract patients from the bus and triage them into categories by color or priority
    C. Wait until an incident commander arrives on scene and then follow his or her direction
    D. Assume command of the incident and give a preliminary report to dispatch

68. Which of the following is not a responsibility of the hazardous materials first responder?

    A. The ability to recognize the need for specialty resources
    B. Entry to the hot zone and mitigation of the incident
    C. Knowledge of hazardous materials and the risks associated with them in case of an accident
    D. An understanding of the potential outcomes of a hazardous materials emergency

69. All of the following are symptoms left ventricular heart failure except:

    A. Paroxysmal nocturnal dyspnea
    B. Orthopnea
    C. Headache
    D. Dyspnea

70. All of the following are stages of cardiac muscle excitation except:

    A. Polarization
    B. Repolarization
    C. Depolarization
    D. Reincarnation

71. In a trauma patient, distended neck veins are most likely a sign of:

    A. Tension pneumothorax
    B. Cervical spine injury
    C. Hemothorax
    D. Profound shock

72. "Crackles" are defined as:

    A. A gurgling sound
    B. A rattling sound
    C. A dry rubbing sound
    D. A fine, bubbling sound

73. The immediate care for almost any chemical burn includes:

    A. Irrigating it with vinegar
    B. Irrigating it with cool or cold water
    C. Scrubbing it with hot water and soap
    D. Covering with a dry sterile dressing

74. Signs and symptoms of renal failure include all of the following except:

    A. Polyuria
    B. Jaundice
    C. Peripheral and peritoneal edema
    D. Cardiac dysrhythmias

75. The primary drug used to treat angina in the field after oxygen is:

    A. Procardia
    B. Atropine
    C. Nitroglycerin
    D. Epinephrine

76. What is the normal pH range of the body?

    A. 7.45 – 7.55
    B. 7.35 – 7.45
    C. 7.15 – 7.25
    D. 7.25 – 7.35

77. What type of injury does lightening cause?

    A. Neuropsychologic problems
    B. Cardiac asystole or ventricular fibrillation
    C. Loss of consciousness or AMS
    D. All of the above

78. Burns are the leading cause of trauma in the _____ age group.

   A. Toddler and pre-school
   B. Elderly
   C. Newborn
   D. Teenage

79. The "rule of nines" is a method of determining:

   A. The type and degree of the burn
   B. The severity of the burn
   C. The classification of the burn
   D. The body surface area burned

80. The semi-lunar valves include the _____valves:

   A. Tricuspid and aortic
   B. Mitral and tricuspid
   C. Pulmonic and aortic
   D. Pulmonic and mitral

81. Examples of factors that may change the scene time include:

   A. Heavy traffic conditions
   B. A change in transport destination
   C. A lengthy extrication
   D. The dispatch priority

82. Upon arrival at the scene of a MVC, you find that the driver is slumped over the steering wheel. You manually stabilize her cervical spine and see that she struck her neck on the rim of the steering wheel. Which of the following airway problems can you expect with this patient?

    A. Laryngeal spasm
    B. Laryngeal edema
    C. Fractured larynx
    D. All of the above

83. The injection of a drug into the spinal canal on or outside the dura mater that surrounds the spinal column is called a (an):

    A. Intravenous
    B. Epidural
    C. Subdural
    D. Intrapleural

84. Acquired hypersensitivity is called a (an):

    A. Allergy
    B. Cancer
    C. Rheumatic fever
    D. Asthma

85. A state EMS act is an example of _____ law.

    A. Administrative
    B. Civil
    C. Legislative
    D. Criminal

86. At one year after the cessation of cigarette smoking, the excess risk
     of coronary heart diseases is decreased to _____ of a smoker:

     A. Two-thirds
     B. One-fourth
     C. One-third
     D. One-half

87. The orbits of the eye (eye sockets) are formed by:

     A. The maxillary and nasal bones
     B. The frontal bones
     C. The zygomatic bones
     D. All of the above

88. The sacrum consists of _____ vertebrae that are fused together:

     A. 12
     B. 5
     C. 7
     D. 3 to 5

89. The body's thermo-regulatory control mechanism is located in the:

     A. Cerebrum
     B. Pons
     C. Hypothalamus
     D. Medulla oblongata

90.   Which of the following is the best definition of "Cardiac tamponade"?

    A.   As blood fills the pericardium, the aortic and mitral valves become narrowed

    B.   As blood fills the pericardium, the blood pressure elevates dramatically

    C.   As blood fills the pericardial sac, the coronary arteries become occluded

    D.   As blood fills the pericardium, the heart's function is progressively compromised

91.   All of the following are possible causes of infectious and/ or communicable diseases except:

    A.   Viruses

    B.   Carcinoma

    C.   Bacteria

    D.   Parasites

92.   All of the following are categories of Vascular access devices used in home health care except:

    A.   Implanted ports

    B.   Peripheral inserted central catheters

    C.   Central venous catheters

    D.   Arterial lines

93.   Smallpox, Marburg, and Lassa fever are all caused by a _____.

    A.   Viral infection

    B.   Bacterial infection

    C.   Man-made organisms

    D.   Enterotoxin

94.   What is another term for a cumulative stress reaction?

      A.   Anxiety
      B.   Fight or flight reaction
      C.   Flashback
      D.   Burnout

95.   Moving the outstretched forearm so that the anterior surface is facing downward is called:

      A.   Extension
      B.   Rotation
      C.   Pronation
      D.   Supination

96.   The two primary goals of pre-hospital care of a sexual assault victim are to preserve the victim's privacy and dignity and to:

      A.   Help the victim bathe and change her clothing
      B.   Preserve all physical evidence for the police
      C.   Obtain a complete description of the assailant
      D.   Collect samples of body fluids from the patient

97.   What does it mean if a woman is described as Multipara?

      A.   She is over 45 years old and pregnant
      B.   She is pregnant and morbidly obese
      C.   She has never delivered an viable infant
      D.   She has delivered more that one baby

98.   Which of the following is an example of acknowledging and labeling the patient's feelings?

      A.   "Anger is a hostile emotion. Let's be more positive"
      B.   "Stop threatening me. I've never hurt you."
      C.   "You seem angry. Do you want to tell me about it?"
      D.   "I get angry myself sometimes"

99.   Of the following, which facial bone is the most frequently fractured?

      A.   Mandible
      B.   Nose
      C.   Maxilla
      D.   Zygoma

100. The ethmoid bone lies:

      A.   Beyond the frontal bone
      B.   In the nasal bone
      C.   In the center of the cranium
      D.   In the middle of the neck

101. The optic nerve is cranial nerve number:

      A.   IV
      B.   X
      C.   III
      D.   II

102. At birth, an infants total body weight (TBW) is approximately:

    A. 90 – 95% of the TBW
    B. 75 – 80% of the TBW
    C. 20 – 30% of the TBW
    D. 30 – 40% of the TBW

103. The medical term for an excessive urine output is:

    A. Polygastria
    B. Polyhydruria
    C. Polyuria
    D. Polydipsia

104. Medications that are given through the skin by absorption are:

    A. Transtracheal
    B. Transluminar
    C. Transcutaneous
    D. Transdermal

105. If a patient has a head injury, the paramedic must suspect
    _____injuries.

    A. Spinal
    B. Back
    C. Ear
    D. Eye

106. A patient is found in asystole, the most appropriate and immediate intervention is:

    A. Defibrillate at 200 – 3000 joules
    B. Start CPR
    C. Administer epinephrine
    D. Pre-cordial thump

107. Social, religious, or personal standards of right or wrong are:

    A. Rules
    B. Ethics
    C. Morals
    D. Professionalism

108. The placenta performs all of the following functions, except:

    A. Transport of nutrients and excretion of wastes
    B. Transfer of gases (Fetal respiration)
    C. Hormone production
    D. Providing an extra cushion effect to protect the fetus from trauma

109. Safety equipment that EMS personnel use at the scene of a crash includes:

    A. A turnout coat to protect from sharp objects
    B. Head protection, such as a helmet
    C. Eye protection, such as shields and goggles
    D. All of the above

110. Primary responsibilities for returning to service after a call includes all the following, except:

    A. Defriefing
    B. Disinfecting
    C. Regenerating
    D. Restocking

111. Which of the following conditions are covered under the Americans with Disabilities Act (ADA)?

    A. Hearing impairment
    B. Fractured leg
    C. Drug abuse
    D. All of the above

112. Which of the following driving courses provide immunity from lawsuits?

    A. Defensive Driving Course (DDC)
    B. Emergency Vehicle Operator Course (EVOC)
    C. Certified Emergency Vehicle Operator (CEVO)
    D. None of the above

113. The best way for a paramedic to avoid a lawsuit is to:

    A. Accurately document the assessment and management of the patient clearly on the PCR
    B. Always be respectful and pleasant to patient's, their families, their property
    C. Practice good medicine and be a competent caregiver
    D. All of the above

114.   A sex-linked hereditary disorder most commonly passed on from an asymptomatic mother to a male child is:

       A. Hemophilia
       B. Hematochromatosis
       C. Anemia
       D. ALS

115.   When a drug is administered to be dissolved between the cheek and gum, what type of route is this?

       A. Intra-atricular
       B. Buccal
       C. Sublingual
       D. Ingestion

116.   Who should pronounce death?

       A. The EMT- basic on scene
       B. Whoever the state authorizes
       C. The paramedic on scene
       D. All of the above

117.   Like many professions, _____ is an integral component of the EMS profession.

       A. Communication
       B. Diet
       C. Encoding
       D. Sign language

118.   To listen for vesicular breath sounds, place the diaphragm over the:

   A. Fifth posterior axillary line and mid-clavicular line
   B. Second intercostals space at the mid-clavicular line
   C. Fifth anterior axillary line and mid-axillary line at the fifth rib level
   D. Second intercostals space at the mid-axillary line

119.   The frequency of performing ongoing assessments is usually based on:

   A. Medical versus trauma patient
   B. Past medical history
   C. Good clinical judgement and experience
   D. Age of patient

120.   The most common causes of syncope are vasovagal faint, positional orthostatic hypotension and:

   A. Dehydration
   B. Cardiac dysrhythmias
   C. Micturition
   D. Neurologic induced

121.   Which of the following effects from the medication Dopamine is not desired in anaphylaxis?

   A. Maintenance of systolic pressure
   B. Increased cardiac contractibility
   C. Increased peripheral vasoconstriction
   D. Renal and mesentery artery vasodilation

122. Normal wound healing can be altered by which of the following factors?

    A. Dry skin
    B. Skin temperature
    C. Ambient temperature
    D. Body region

123. Soft tissue trauma included all of the following, except:

    A. Spasms and ticks
    B. Bumps and bruises
    C. Hematomas
    D. Lacerations

124. Which of the following is not a major vein of the chest?

    A. Subclavian
    B. Aorta
    C. Internal jugular
    D. External jugular

125. Ribs _____ are most often fractured because they are thin and poorly protected.

    A. 4 to 9
    B. 10 to 12
    C. 1 to 3
    D. None of the above

ANSWER SHEET 4

1.   B- 5,2,4,3,1
2.   A- Unable to cope
3.   C- Histamine
4.   B- Tonic
5.   C- The patient can still talk
6.   A- 10
7.   C- Histamine
8.   A- Criminal
9.   B- Monitor EKG
10. C- Melting and boiling point
11. B- Epiglottitis
12. D- All of the above
13. A- Getting everyone in the vehicle to wear seat belts
14. B- 0.44
15. D- Any of the above are correct
16. D- Weakening of the chest wall
17. D- It is a surgical procedure
18. B- Encoding
19. D- Significant MOI with neck injury
20. A- Provide some legally defensible backup for the paramedic
21. D- Frontal lobe
22. B- Small piece of stool
23. D- SA node
24. A- Increase the myocardial oxygen consumption
25. A- True
26. D- The occupants collide with the interior of the vehicle
27. D- Cramp-like pain in the calf
28. B- Respiratory depression and constricted pupils
29. D- Nitroglycerin
30. D- The onset of a dysrhythmia
31. D- 18

32. A- SA node
33. B- Oxygen
34. C- Release one corner of the occlusive dressing
35. D- Soil bacterium
36. A- The end of the third stage of labor
37. B- Diphenhydramine
38. C- 10 – 50 mg IV
39. D- Both B and C
40. B- The minimum effective concentration
41. D- Tolerance
42. B- Atropine
43. D- Hyperglycemia
44. A- Foot-pounds
45. C- 15 cc
46. D- Males
47. A- Metabolic acidosis
48. C- 0.1 mg/kg via IV bolus
49. A- Immobilization and elevation of the extremity
50. D- Both A and C

ANSWER SHEET 4

51. D- All of the above
52. C- A wet sterile dressing and an occlusive dressing
53. D- Mood disorder
54. A- Explain what they are going to do
55. B- Lie to the patient
56. C- 1,000
57. A- Nortriptyline
58. C- SA node
59. C- 8
60. A- Schedule II drugs
61. D- Systole
62. D- It depends on the severity of the burn
63. D- Pneumothorax
64. A- Multiplex
65. A- Ruptured spleen
66. A- Sacral
67. D- Assume command of the incident and give a preliminary report to dispatch
68. B- Entry to the hot zone and mitigation of the incident
69. C- Headache
70. D- Reincarnation
71. A- Tension pneumothorax
72. D- A fine, bubbling sound
73. B- Irrigating it with cool or cold water
74. D- Cardiac dysrhythmias
75. C- Nitroglycerin
76. B- 7.35 – 7.45
77. D- All of the above
78. A- Toddler and pre-school
79. D- The body surface area burned
80. C- Pulmonic and aortic

81. C- A lengthy extrication
82. D- All of the above
83. B- Epidural
84. A- Allergy
85. C- Legislative
86. D- One-half
87. D- All of the above
88. B- 5
89. C- Hypothalamus
90. D- As blood fills the pericardium, the heart's function is progressively compromised
91. B- Carcinoma
92. D- Arterial lines
93. A- Viral infection
94. D- Burnout
95. C- Pronation
96. B- Preserve all physical evidence for the police
97. D- She has delivered more than one baby
98. C- "You seem angry. Do you want to tell me about it?"
99. B- Nose
100. B- In the nasal bone

ANSWER SHEET 4

101. D- II
102. B- 75 – 80% of the TBW
103. C- Polyuria
104. D- Transdermal
105. A- Spinal
106. B- Start CPR
107. C- Morals
108. D- Providing an extra cushion effect to protect the fetus from trauma
109. D- All of the above
110. C- Regenerating
111. A- Hearing impairment
112. D- None of the above
113. D- All of the above
114. A- Hemophilia
115. B- Buccal
116. B- Whoever the state authorizes
117. A- Communication
118. C- Fifth anterior axillary line and and mid-axillary linge at the fifth rib level
119. C- Good clinical judgement and experience
120. B- Cardiac dysrhythmias
121. D- Renal and mesentery artery vasodilation
122. D- Body region
123. A- Spasms and ticks
124. B- Aorta
125. A- 4 – 9

Test 5

1. The drug that is indicated in pre-eclampsia and eclampsia and that may be order by medical control is:

    A. Oxytoxin
    B. Magnesium sulfate
    C. Pitocin
    D. Calcium chloride

2. The birth of the baby signifies:

    A. The end of the second stage of labor
    B. The beginning of the fourth stage of labor
    C. The end of the first stage of labor
    D. The beginning of the third stage of labor

3. Procainamide is considered a _____channel blocker medication:

    A. Dopaminergic
    B. Beta
    C. Sodium
    D. Calcium

4. The antidotes for organophosphate poisoning are:

    A. Epinephrine and atropine
    B. Mucomyst and sodium bicarbonate
    C. Atropine and pralidoxamine
    D. Lidocaine and dopamine

5. Most patients responding to psychological stress in an abnormal way are:

    A. Of no real danger to themselves
    B. Of no real danger to others
    C. Very dangerous
    D. Both A and B

6. You are dispatched to a grocery store for a 46 year old unconscious patient. However, when you arrive and begin to examine the patient, you note that he is unresponsive, pulseless, and apneic. You immediately reach for which of the following piece of equipment?

    A. Quick look monitor defibrillator paddles
    B. Oxygen mask
    C. IV dextrose
    D. IV line

7. All of the following are possible nervous system presentations of chronic renal failure except:

    A. Muscle twitching
    B. Tinnitus
    C. Delirium
    D. Seizures

8. All of the following are parts of the hematopoietic (blood cell producing) system except:

    A. Spleen
    B. Liver
    C. Yellow bone marrow
    D. Red bone marrow

9. Which of the following is the best definition of "high altitude sickness"?

    A. On exposure to reduced atmospheric pressures, hypobaric hypoxia occurs
    B. On exposure to reduced atmospheric pressures, the patient begins to feel nervous
    C. On flying in an airplane, on ascent, the patient feels his or her ears pop
    D. On walking up a mountain, the patient becomes anxious

10. The posterior tibial pulse can be palpated near the:

    A. Top of the foot
    B. Medial ankle bone
    C. Arch of the foot
    D. Posterior knee

11. What is the primary drug for the management of acute anaphylaxis?

    A. Terbutaline
    B. Diphenhydramine HCL
    C. Methylprednisolone
    D. Epinephrine

12. A greenstick fracture is one that is:

    A. Comminuted
    B. Partial
    C. Open
    D. Impacted

13. During delivery, you notice that the amniotic fluid is discolored and has a foul odor. What should you do first?

    A. Provide five back blows and then five chest thrusts
    B. Intubate the child and give positive pressure ventilations
    C. Suction the upper airway, using a meconium aspirator
    D. Dry, warm, position, suction, and stimulate the child to breath

14. One of the best ways for EMS personnel to deal with job-related stress is to:

    A. Eliminate all physical exercises
    B. Discuss the situation with co-workers
    C. Take sleeping pills at night as needed
    D. Take time away from family and friends

15. An elderly female is complaining of a sudden onset of severe pain in her left leg. The affected extremity is cool to the touch and pale. The temperature and pulse in the patient's right leg is normal. You suspect:

    A. Deep vein thrombosis
    B. Varicose arteries
    C. Pulmonary embolism
    D. Arterial occlusion

16. What is an assessment finding of pulsus paradoxus associated with?

    A. Myocardial infarction
    B. Emphysema
    C. COPD
    D. Congestive heart disease

17. Written procedures for the management of pre-hospital emergencies that are approved by representatives of the medical community are known as:

    A. Protocols
    B. Conditions of employment
    C. Online medical control
    D. DOT standards

18. When using the OPQRST mnemonic to assess the patient's pain, you would assess the R portion of the mnemonic by asking:

    A. Does the pain feel sharp or dull?
    B. When did it start hurting you?
    C. What makes it feel better?
    D. Does the pain move anywhere?

19. Which of the following patient presentations would best be managed by a nasal intubation?

    A. The patient is unconscious, is breathing slowly, and has a gag reflex
    B. The patient is semiconscious, is breathing slowly, and has a pulse
    C. The patient is unconscious, and apneic with a pulse
    D. The patient is unconscious, apneic, and pulseless

20.   Uterine inversion is when the uterus literally turns inside out upon delivery of the infant and/ or placenta. Uterine inversion:

    A.   Produces profound, life threatening shock
    B.   Can be caused by pulling on the umbilical cord
    C.   Can be caused by attempts to express the placenta when the uterus is relaxed
    D.   All of the above

21.   Suppositories should NEVER be administered when:

    A.   Hemorrhoids are present
    B.   Presence of diarrhea
    C.   There is evidence of rectal bleeding
    D.   All of the above

22.   Signs and symptoms of a patient experiencing a CVA or TIA include all of the following except:

    A.   Vision disturbances
    B.   Unresponsiveness
    C.   Paraparesis or paraplegia
    D.   Alter level of consciousness, confusion, or agitation

23.   Each nerve cell has branches that receive impulses and carry them to the cell body. Each of these branches is called:

    A.   A dendrite
    B.   An axon
    C.   A ganglia
    D.   A neuron

24. Water is the universal _____ and sodium chloride is the universal:

    A. Salt, solution
    B. Solvent, solute
    C. Solution, solute
    D. Solute, solvent

25. An accumulation of air in the pleural space of the lungs is called:

    A. Hemithorax
    B. Hemothorax
    C. Pneumothorax
    D. Laterothorax

26. EMS research is important for the growth and evolution of the profession because it:

    A. Strengthens pre-hospital treatment modalities
    B. Helps build a foundation
    C. Establishes credibility
    D. All of the above

27. Balloon like sacs at the end of the bronchioles where gas exchange takes place:

    A. Distributive sacs
    B. Alveoli
    C. Bronchus
    D. Air sacs

28. When a paramedic documents a safety hazard, which of the following should be included:

    A. Information required by the EMS agency
    B. Primary care provided
    C. Primary injury data
    D. All of the above

29. In _____the first know air medical transport occurred during the retreat of the Serbian army from Albania:

    A. 1962
    B. 1955
    C. 1930
    D. 1915

30. All of the following immunizations are required for paramedics, except:

    A. Hepatitis B
    B. Lyme disease
    C. Tetanus and diphtheria
    D. Rubella

31. The leading cause of chronic illness in children is caused by:

    A. Asthma
    B. Diabetes
    C. Chicken pox
    D. Cancer

32. One of the primary advantages for giving drugs by parenteral route in emergencies is that:

    A. Absorption effects are more predictable
    B. IV access is always available
    C. It is the most economical
    D. It is convenient

33. The presence of ascites upon physical examination is an abnormal finding on which area of the body?

    A. Extremities
    B. Head
    C. Neck
    D. Abdomen

34. The bell of the stethoscope is best used for listening to _____ sounds.

    A. Breath
    B. Abnormal heart
    C. Normal heart
    D. Bowel

35. The paramedic can improve scene choreography by:

    A. Taking a course in MCI leadership
    B. Watching training videos
    C. Preplanning and practice with his/ her crew
    D. Utilizing distance learning

36.   The load against which the heart exerts its contractile force is the:

      A.   Overload
      B.   Preload
      C.   Frontload
      D.   Afterload

37.   Which of the following is a life-threatening condition with a pathology of an accumulation of fluids into the pericardial sac?

      A.   Cardiac tamponade
      B.   Myocarditis
      C.   Pericarditis
      D.   Pneumopericardium

38.   Peripheral edema is more likely to be present with _____ because it takes several hours to develop?

      A.   Bronchitis
      B.   Chronic CHF
      C.   Pneumonia
      D.   Acute myocardial infarction

39.   The fundamental component of the nervous system is:

      A.   Nerve impulse
      B.   Myelin
      C.   Neuron
      D.   Spinal cord

40. The twelve pair of cranial nerves are a component of which part of the nervous system?

    A. Functional
    B. Autonomic
    C. Peripheral
    D. Structural

41. Conditions that result in abdominal pain, but do not originate in the abdomen, include all the following, except:

    A. Food poisoning
    B. Pneumonia
    C. Herpes zoster
    D. Black widow spider bite

42. The Ryan-White Act of 1990 requires that exposure notification to emergency responders must by made within _____ hours.

    A. 72
    B. 24
    C. 48
    D. 12

43. What is the most common symptom of active TB?

    A. Weakness
    B. Productive cough
    C. Shortness of breath
    D. Fever

44. When a projectile such as a bullet passes through the body, it creates a wave of pressure that can compress organs and tissue causing:

    A. Spontaneous combustion
    B. Collapse and disintegration
    C. Liquidation
    D. Contusion, fracture or rupture

45. The victim of a motorcycle collision that does not wear a helmet has a _____% increased risk of brain injury.

    A. 500
    B. 300
    C. 30
    D. 50

46. Abdominal trauma is the _____ leading cause of preventable trauma death.

    A. Fifth
    B. Third
    C. Second
    D. Fourth

47. As the body ages, the skin begins to sag and wrinkles develop because of:

    A. Loss of sebaceous glands
    B. Loss of elastic fiber
    C. Increased vascularity in the skin
    D. Loss of T cell function

48. If it is necessary to lift a patient vertically out of a narrow hole, the best device to consider using is a:

     A. SKED
     B. Stokes basket
     C. KED
     D. Scoop stretcher

49. What safety precautions should be taken at all rescue scenes?

     A. Be alert to traffic at the scene
     B. Ensure all rescuers wear PPE
     C. Make sure EMS personnel are visible
     D. All of the above

50. If a person is found hanging and the paramedic is going to attempt a resuscitation, the paramedic should wear gloves and take the patient down by _____ the knot.

     A. Cutting to avoid
     B. Cutting through
     C. Getting the police to cut
     D. Untying

51. Overdose of tricyclic antidepressants has been shown to cause hypotension or:

     A. Hypothermia
     B. Ventricular dysrhythmias
     C. Hypomagnesium
     D. TIA or stroke

52. The three phases associated with the blast effect are the primary phase, the secondary phase, and the _____ phase.

    A. End-stage
    B. Late
    C. Triage
    D. Tertiary

53. During the secondary phase of the blast effect, the potential for injury comes from:

    A. The patient striking an object
    B. Pressure waves
    C. Flying articles
    D. The heat wave

54. This structure lies in the midline of the neck and is inferior to the hyoid bones and anterior to the esophagus:

    A. Uvula
    B. Larynx
    C. Pharynx
    D. Trachea

55. The most abundant tissue in the body is:

    A. Nerve tissue
    B. Endothelial tissue
    C. Connective tissue
    D. Skeletal muscle tissue

56. A disadvantage of the Combi-tube is:

   A. Esophageal involvement
   B. Blind insertion
   C. Too large
   D. Both A & B are correct

57. A statement which addresses humanitarian concerns in regards to a respective profession is:

   A. Code of ethics
   B. Rules
   C. Law
   D. Code of conduct

58. Three types of muscle tissue include:

   A. Skeletal, dorsal, cardiac
   B. Skeletal, cardiac, smooth
   C. Cardiac, pectoral, smooth
   D. Pectoral, ventral, dorsal

59. Kussmaul's Respirations are frequently associated with:

   A. Hypoglycemia
   B. Hyperglycemia or diabetic ketoacidosis
   C. Hyperglycemia
   D. Hypoglycemia or diabetic ketoacidosis

60. If chemical cardioversion is warranted, which of the following drugs is used initially to treat symptomatic PSVT?

    A. Propanolol
    B. Adenocard
    C. Verapamil
    D. Inderal

61. The glass container which must be broken to obtain the within is called:

    A. Suspension
    B. Vial
    C. Capsule
    D. Ampule

62. A 10% solution would have 10 mg of a solute material in what volume of fluid?

    A. 1 ml
    B. 1000 ml
    C. 100 ml
    D. 10 ml

63. A wound that has a special danger of infection due to anaerobic bacteria is the:

    A. Avulsion
    B. Puncture
    C. Incision
    D. Contusion

64. Spinal fracture occurs most commonly in the:

    A. Lumbar and sacral spine
    B. Cervical and thoracic spine
    C. Thoracic and lumbar spine
    D. Cervical and lumbar spine

65. Mannitol may be given to the patient with a head injury who exhibits:

    A. Drop in blood pressure
    B. Unconsciousness
    C. Signs of an increase in intracranial pressure
    D. Life threatening cardiac dysrhythmias

66. The most noticeable veins distended in congestive heart failure are the:

    A. Venae cava
    B. Pulmonary
    C. Jugular
    D. Cephalic

67. If you have 200 mg of Dopamine to add to 250 cc of D5W, what will be the concentration of Dopamine?

    A. 125 mcg/ml
    B. 0.8 mg/ml
    C. 80 mcg/ml
    D. 1.25 mg/ml

68. The term Motor Vehicle accident is now called Motor Vehicle:

   A. Upset
   B. Injury
   C. Threat
   D. Crash

69. The basis for most paramedic training programs is the National curriculum developed by:

   A. The American Red Cross
   B. The American Heart Association
   C. The American Medical Association
   D. The United States Department of Transportation

70. Proper nutrition involves an understanding of nutrients the body needs, as well as the principles of:

   A. Weight control
   B. Mental health
   C. Exercise
   D. Emotional well- being

71. The key techniques for reducing crisis-induced stress include all the following, except:

   A. Limiting exposure to the incident
   B. Getting plenty of rest
   C. Replacing food and fluids
   D. Increased cigarette smoking

72. All of the following are cognitive signs and symptoms of stress, except:

    A. Fear
    B. Distressing dreams
    C. Disorientation
    D. Memory problems

73. Which of the following is a general type of law?

    A. Criminal
    B. Legislative
    C. Administrative
    D. All of the above

74. Minors who can refuse treatment include:

    A. Members of the armed forces
    B. Those who are married
    C. Those who are parents themselves
    D. All of the above

75. The study of abnormal function of the body is called:

    A. Pathophysiology
    B. Physiology
    C. Anatomy
    D. Pathology

76.   The ability of microorganisms to cause disease is called:

       A. Infectability
       B. Virulence
       C. Toxicity
       D. Carcinogenesis

77.   A solution with a lower solute concentration than the blood is referred
       to as a/ an _____ solution.

       A. Neotonic
       B. Hypertonic
       C. Hypotonic
       D. Isotonic

78.   A patient may become hypokalemic from:

       A. Excessive vomiting or diarrhea
       B. Decreased sweating
       C. A diet deficient in sodium
       D. Excessive eating

79.   What is MODS?

       A. Advice for managing shock
       B. A disease of the brain
       C. Multiple organ dysfunction syndrome
       D. None of the above

80. Convert 1.5 to percent:

      A. 1.50 %
      B. 1500 %
      C. 150 %
      D. 15 %

81. You are to administer D50W to an unresponsive diabetic patient. The D50W comes in a 500 ml (preload) syringe and you will administer all of it. How many grams of dextrose will you administer?

      A. 500
      B. 50
      C. 5
      D. 25

82. What is 250 % of 30?

      A. 75
      B. 7500
      C. 0.12
      D. 8.33

83. The general term now used to describe the spectrum of disease from acute angina to myocardial infarction is:

      A. Coronary illness
      B. Acute coronary syndrome
      C. Heart attack
      D. Unstable angina

84. Which statement is not correct?

    A. The jaw thrust can be used for one rescuer BVM technique on a trauma patient
    B. The BVM should be attached to 100 % oxygen
    C. Proper use of the BVM requires practice
    D. Tidal volumes of 400 600 ml can be given over one second

85. _____ is a lack of oxygen in the blood.

    A. Hypoxia
    B. Anaerobic metabolism
    C. Hypoxemia
    D. Anoxia

86. _____ is a last resort airway technique when a patient has a complete airway obstruction or in whom tracheal intubation is otherwise impossible.

    A. LMA
    B. Needle decompression
    C. Rapid sequence intubation
    D. Needle cricothyrotomy

87. Which of the following is not a form of communication used in the EMS profession?

    A. Prayer
    B. Written
    C. Verbal
    D. Non-verbal

88. When you are not getting the answer to questions that appear obvious, consider that the patient may be:

    A. Not ill or embarrassed
    B. Pregnant
    C. Embarrassed
    D. Falsely reassured

89. Which of the following is not considered part of a health history:

    A. Current vital signs
    B. Religious beliefs
    C. Daily living activities
    D. Career status

90. An example of a patient's specific characteristic that is apparent on visual inspection is:

    A. Crepitus
    B. Personal hygiene
    C. Wheezing
    D. Bruits

91. A device consisting of a light source and several lenses that is used to look at the eye is called a/ an:

    A. Doppler
    B. Otoscope
    C. Penlight
    D. Ophthalmoscope

92. _____ is an irregular body texture characterized by air trapped under the skin:

    A. Tenting
    B. Edema
    C. Subcutaneous emphysema
    D. Crepitus

93. _____ is a term used when assessing a patient's skin for state of hydration:

    A. Turgor
    B. Rigidity
    C. Hydrolysis
    D. Edema

94. The components of the paramedic's patient assessment begin with the:

    A. General impression
    B. Scene size-up
    C. Initial assessment
    D. Focused history

95. The standard assessment is the _____ format:

    A. Head to toe
    B. Vectored exam
    C. Toe to head
    D. Focused exam

96. To evaluate the mental status of an unconscious patient, check for deep pain response and:

 A. Pulse
 B. Glucose
 C. Reflexes
 D. EKG

97. Good documentation of the patient on the run report is part of the _____component of patient assessment:

 A. Protocol
 B. Legal
 C. Moral
 D. Ethical

98. The second component of an assessment in the field is:

 A. Initial assessment
 B. Provide high flow oxygen
 C. Scene size up
 D. Open the airway

99. While assessing a patient for a painful response, which of the following is an appropriate response?

 A. Flexing the arm
 B. No response
 C. Withdraw from pain
 D. Extending the arm

100. Cardiac patient are often medicated to prevent vomiting for the following reason:

    A. Hypoxia
    B. To prevent discomfort
    C. Dehydration
    D. Vagus stimulation

101. The _____is a numeric grading system that combines the GCS and measurements of cardiopulmonary function as a gauge of the severity of injury and a predictor of survival after blunt injury to the head:

    A. Trauma score
    B. TBI scale
    C. Cincinnati score
    D. CUPS

102. All of the following are examples of stimulants of the fight or flight response for paramedics, except:

    A. Traffic
    B. Lights and sirens
    C. Nitro spray
    D. Pagers

103. The basic model of communications includes each of the following steps, except:

    A. Receiver gives feedback
    B. Sender decodes the message
    C. Sender sends the message
    D. Receiver receives the message

104.   The suffix "Dynia" means:

A.   Through
B.   Painful condition
C.   To free
D.   Outer

105.   Which of the following is an aspect of professionalism?

A.   Maintaining patient confidentiality
B.   Attend regular continuing educational programs
C.   Being well-groomed
D.   All of the above

106.   Every paramedic MUST get fitted for proper face seal to utilize a
_____ mask.

A.   PTCA
B.   HEPA
C.   BIPI
D.   PWCP

107.   Back injuries account for _____% of work place injuries:

A.   80 %
B.   22 %
C.   50 %
D.   30 %

108. The Sawtooth wave is characteristic of the following dysrhythmias?

    A. Atrial flutter
    B. Ventricular bigeminy
    C. Atrial fibrillation
    D. Ventricular tachycardia

109. The normal P – R interval is about:

    A. .20 - .24 seconds
    B. .05 - .10 seconds
    C. .12 - .20 seconds
    D. .18 - .24 seconds

110. The most common chief complaint in patients with heart disease is:

    A. Chest pain
    B. Syncope
    C. Dyspnea
    D. Palpitations

111. The baseline of the EKG is referred to as the:

    A. Clothes line
    B. Isoelectric line
    C. Reference point line
    D. Neutral line

112.   Prior to beginning practice as a Paramedic, it is essential that you
        acquire which of the following?

        A. County certification
        B. State certification and / or licensure
        C. National licensure
        D. Paramedic course completion

113.   The Paramedic may be required to report all of the following types
        of cases to the proper authorities except:

        A. Child abuse
        B. Rape
        C. Stab wounds
        D. Myocardial infarction in a 48 year old female

114.   The most important source for obtaining information on the
        patient's history is:

        A. The patient's personal physician
        B. The patient
        C. The patient's family
        D. Witnesses at the scene

115.   In relationship to the fracture site, the Paramedic should check the
        _____pulse:

        A. Distal
        B. Proximal
        C. Lateral
        D. Medial

116.   A type of open wound with jagged edges caused by tearing forces is known as:

      A.   A puncture
      B.   An abrasion
      C.   A laceration
      D.   An incision

117.   An injury where soft tissue has been torn away is known as:

      A.   A contusion
      B.   An avulsion
      C.   An amputation
      D.   An abrasion

118.   Of the following, which IV solution set would provide the most rapid fluid infusion?

      A.   Burette
      B.   60 gtts/ ml
      C.   10 gtts/ ml
      D.   15 gtts/ ml

119.   Define a mass casualty incident:

      A.   An incident that overwhelms local resources
      B.   A terrorist event
      C.   An incident involving multiple response agencies
      D.   An incident that can be managed by the initial responding unit

120.   A patient has been stabbed in the back. Which of the following signs would most likely make you suspect that the patient has a kidney injury?

        A. Thirst
        B. Hematuria
        C. Abdominal tenderness
        D. Ecchymosis to the flank

121.   When is use of "reasonable" force or use of restraints permissible for a patient?

        A. It is never appropriate to use any forced on a patient
        B. This is at the discretion of the paramedic or the crew's officer
        C. The patient is alert, cooperative but anxious
        D. The patient has altered level of consciousness caused by injury, substance abuse, or illness

122.   The three general types of amputations are:

        A. Complete, near-complete and degloving
        B. Complete, partial and near-complete
        C. Complete, partial and degloving
        D. Complete, degloving and avulsion

123.   Your patient is ready to deliver her second child. She has a one year old daughter and has had one previous miscarriage. You would describe her as being:

        A. Gravida 3, para 1
        B. Para 1, gravida 2
        C. Gravida 2, para 1
        D. Para 2, gravida 3

124. The classic sign& symptom that delivery is imminent is:

     A. Contractions only 5
     B. only 15 to 30 seconds in duration
     C. The rupture of the amniotic sac
     D. The uncontrollable maternal urge to push or the crowning of the fetus

125. Which of the following would not normally trigger a stress reaction?

     A. Death of a loved one
     B. Personal injury
     C. Complimentary letter
     D. Job stress

ANSWER SHEET 5

1.   B- Magnesium sulfate
2.   A- The beginning of the second stage of labor
3.   C- Sodium
4.   C- Atropine and pralidoxamine
5.   C- Very dangerous
6.   A- Quick look monitor defibrillator paddles
7.   B- Tinnitus
8.   C- Yellow bone marrow
9.   A- On exposure to reduced atmospheric pressures, hypobaric hypoxia
      occurs
10.   B- Medial ankle bone
11.   D- Epinephrine
12.   B- Partial
13.   C- Suction the upper airway, using a meconium aspirator
14.   B- Discuss the situation with co-workers
15.   D- Arterial occlusion
16.   C- COPD
17.   A- Protocols
18.   D- Does the pain move anywhere?
19.   A- The patient is unconscious, is breathing slowly, and has gag reflex
20.   D- All of the above
21.   D- All of the above
22.   C- Paraparesis or paraplegia
23.   A- A dendrite
24.   B- Solvent, solute
25.   C- Pneumthorax
26.   D- All of the above
27.   B- Alveoli
28.   D- All of the above
29.   D- 1915
30.   B- Lyme disease

31. A- Asthma
32. A- Absorption effects are more predictable
33. D- Abdomen
34. B- Abnormal heart
35. C- Preplanning and practice with his/ her crew
36. D- Afterload
37. A- Cardiac tamponade
38. B- Chronic CHF
39. C- Neuron
40. B- Autonomic
41. A- Food poisoning
42. C- 48
43. B- Productive cough
44. D- Contusion, fracture, or rupture
45. B- 300
46. C- Second
47. B- Loss of elastic fiber
48. A- SKED
49. D- All of the above
50. A- Cutting to avoid

ANSWER SHEET 5

51. B- Ventricular dysrhythmias
52. D- Tertiary
53. C- Flying articles
54. B- Larynx
55. C- Connective tissure
56. D- Both A & B are correct
57. A- Code of ethics
58. B- Skeletal, cardiac, smooth
59. A- Hypoglycemia
60. B- Adenocard
61. D- Ampule
62. C- 100 ml
63. B- Puncture
64. D- Cervical and lumbar spine
65. C- Signs of an increase in intracranial pressure
66. C- Jugular
67. B- 0.8 mg/ml
68. D- Crash
69. D- The United States Department of Transportation
70. A- Weight control
71. D- Increased cigarette smoking
72. A- Fear
73. D- All of the above
74. D- All of the above
75. A- Pathophysiology
76. B- Virulence
77. C- Hypontonic
78. A- Excessive vomiting or diarrhea
79. C- Multiple organ dysfunction syndrome
80. C- 150%
81. D- 25

82. A- 75
83. B- Acute coronary syndrome
84. A- The jaw thrust can be used for one rescuer BVM technique on a trauma patient
85. C- Hypoxemia
86. D- Needle cricothyrotomy
87. A- Prayer
88. C- Embarrassed
89. A- Current vital signs
90. B- Personal hygiene
91. D- Ophthalmoscope
92. C- Subcutaneous emphysema
93. A- Turgor
94. B- Scene size-up
95. A- Head to toe
96. C- Reflexes
97. B- Legal
98. A- Initial assessment
99. C- Withdraw from pain
100. D- Vagus stimulation

ANSWER SHEET 5

101. A- Trauma score
102. C- Nitro spray
103. B- Sender decodes the message
104. B- Painful condition
105. D- All of the above
106. B- HEPA
107. B- 22%
108. A- Atrial flutter
109. C- .12 - .20 seconds
110. A- Chest pain
111. B- Isoelectric line
112. B- State certification and / or licensure
113. D- Myocardial infarction in a 48 year old female\
114. B- The patient
115. A- Distal
116. C- A laceration
117. D- An abrasion
118. C- 10 gtts/ ml
119. A- An incident that overwhelms local resources
120. B- Hematuria
121. D- The patient has altered level of consciousness caused by injury, substance abuse, or illness
122. C- Complete, partial and degloving
123. A- Gravida 3, para 1
124. D- The uncontrollable maternal urge to push or the crowning of the fetus
125. C- Complimentary letter

1. Drugs that are used to reduce fever and interrupt the inflammatory process are:

    A. NIDDMS
    B. NSAIDS
    C. MAO Inhibitors
    D. ACE Inhibitors

2. Drugs are either unchanged or metabolized prior to elimination. The body eliminated drugs in all of the following ways but one:

    A. Sweat
    B. Urine
    C. Feces
    D. Expired air

3. Which of the following statements concerning near-drowning is INCORRECT?

    A. Approximately 40% of near-drowning victims do not aspirate water
    B. Most adult drownings are associated with alcohol
    C. Any near drowning patient should be transported to the hospital
    D. Prolonged survival is possible with cold water immersion

4. When the anterior neck is palpated from the top, downwards, the cricoid cartilage is:

    A. Between the first and second prominent structures
    B. The second prominent structure felt
    C. The first prominent structure felt
    D. The third prominent structure felt

5. Laryngeal spasms may occur secondary to:

    A. Bronchospasm
    B. Spinal trauma
    C. Edema of the glottis
    D. None of the above

6. Cells found in the bone marrow from which all blood cells come:

    A. Stem cells
    B. Growth cells
    C. Mast cells
    D. Chief cells

7. All of the following are stages of the grieving process except:

    A. Depression
    B. Denial
    C. Anger
    D. Exhilaration

8. Of the following, which would be the least likely to be used as a defense mechanism that the Paramedic would use to deal with a stressful event?

    A. Isolation
    B. Crying
    C. Denial
    D. Rationalization

9. A controlled substance that is classified as schedule I is defined as a drug with:

    A. High abuse potential and no medical uses
    B. Low abuse potential and accepted medical uses
    C. High abuse potential and accepted medical uses
    D. Moderate abuse potential and accepted medical uses

10. Which is straighter, right or left main stem bronchi?

    A. Left
    B. Right

11. A 50 year old male is found in pulmonary edema. He is sitting upright in a chair, laboring to breathe. You would expect his P02 to be:

    A. Normal
    B. Low
    C. High

12. Infants and toddlers are best intubated with which type of laryngoscope blade?

    A. Miller
    B. Fiber optic
    C. MacIntosh
    D. Curved

13. Which of the following is an example of an "open-ended" question?

    A. Are you under the care of a physician?
    B. Do you have any allergies?
    C. What type of pain are you experiencing?
    D. Have you been hospitalized recently?

14. All of the following pulse points are generally accepted in the determination of a pulse rate in an adult except:

    A. Brachial
    B. Carotid
    C. Femoral
    D. Radial

15. You are treating a patient with an altered mental status. She is confused and disoriented. With painful stimuli, she opens her eyes and withdraws from the stimuli. What is her score on the Glasgow coma scale?

    A. 12
    B. 9
    C. 10
    D. 11

16. In patients with spinal cord injuries, a loss of sensation below the nipple line would indicate damage at what level of vertebrae?

     A. T1
     B. C5
     C. T4
     D. C6

17. Which of the following fractures should not be straightened?

     A. Tibia
     B. Elbow
     C. Femur
     D. Radius

18. Injuries of the _____ result in referred pain to the left shoulder.

     A. Right lung
     B. Liver
     C. Spleen
     D. Cervical spine

19. A virus that breaks down the body's natural immune defenses is called:

     A. Hepititus C
     B. HIV
     C. Hepititus B
     D. AID

20. The opposite of anabolism is:

   A. Hyperabolism
   B. Catabolism
   C. Proabolism
   D. Antiabolism

21. Water accounts for approximately _____% of the total body weight.

   A. 30%
   B. 50%
   C. 40%
   D. 60%

22. pCO2 is a respiratory:

   A. Acid
   B. Side effect
   C. Basic
   D. Alkali

23. The presence of tachycardia in a patient with respiratory distress may be caused by any of the following except:

   A. Use of bronchodilators
   B. Diaphoresis
   C. Fear
   D. Hypoxemia

24. The most common obstructive airway diseases include asthma and:

   A. Legionnaire's disease
   B. Cystic fibrosis
   C. COPD
   D. URI

25. When one or more valves become narrowed because of congenital damage, the valve is said to be:

   A. Intrinsic
   B. Stenotic
   C. Murmured
   D. Regurgitant

26. You are assessing the pulse rate and quality on a 23 year old female experiencing an asthma attack. You find that the pulse is strong, regular, and tachy; however, the pulse decreases considerably during inspiration. The phenomenon is called pulsus:

   A. Differential
   B. Deficit
   C. Alternans
   D. Paradoxus

27. When there is inadequate blood flow to an organ, such as the heart, _____occurs initially.

   A. Ischemia
   B. Embolus
   C. Infarction
   D. Stroke

28. A patient experiencing a septal wall AMI may have abnormalities in which leads?

    A. I, aVL, V5, and V6
    B. V1 and V2
    C. II, III and aVF
    D. V3, V4

29. When heart muscle becomes hypoxic, the myocardium becomes irritable and may cause:

    A. Dysrhythmias
    B. Systemic ischemia
    C. AMI
    D. Blood loss

30. Nerve cells communicate with each other primarily through the:

    A. Peripheral system
    B. The limbic system
    C. Synapses
    D. Flowing of actyholine

31. The term _____refers to the adjustment of the eyes to variations in distance.

    A. Divergence
    B. Acuity
    C. Accommodation
    D. Conjugate gaze

32. Cushing's triad is a (an) _____ sign of rising ICP.

    A. Late
    B. Not
    C. Early
    D. Unreliable

33. Diabetes is the leading cause of all the following conditions, except:

    A. End-stage kidney failure
    B. Adult blindness
    C. Death
    D. Non-traumatic lower extremity amputations

34. The _____ is considered an organ of both the digestive and the endocrine systems.

    A. Testes
    B. Thyroid
    C. Parathyroid
    D. Pancreas

35. The most common reason a diabetic patient develops DKA is because of:

    A. Infection
    B. Too little insulin
    C. Excess glucagons
    D. Too much insulin

36. Besides epinephrine, what other medication classification does the paramedic administer to a patient in anaphylaxis?

    A. Steroid
    B. Histamine
    C. Vasopressor
    D. Broncodilator

37. _____pain is caused by sudden stretching or distention of a hollow organ.

    A. Radiating
    B. Visceral
    C. Somatic
    D. Biliary

38. All of the following are common causes of lower GI bleeding, except:

    A. Fissures
    B. Tumors
    C. Polyps
    D. Esophageal varices

39. The blood pressure is often elevated with an overdose of:

    A. Pesticides
    B. Cocaine
    C. Depressants
    D. Aspirin

40. A substance may pass through the skin and enter the body. This
     known as:

     A. Absorption
     B. Ingestion
     C. Exhaustion
     D. Inhalation

41. Gastric dialysis is a mechanism of poison removal aided by the
     administration of:

     A. Milk or mild soap
     B. Ipecac syrup
     C. Tincture of benzene
     D. Activated charcoal

42. Body heat is generated as a side effect of normal_____processes.

     A. Neural
     B. Systemic
     C. Metabolic
     D. Cell-mediated

43. Pneumonia is a respiratory disease caused by a:

     A. Fungus
     B. Virus
     C. Bacterial infection
     D. A and B only

44. Cardiac dysrhythmias may accompany a patient suffering from:

    A. Asthma
    B. Emphysema
    C. Chronic bronchitis
    D. All of the above

45. Carbon dioxide is _____ times more soluble in water than oxygen:

    A. 50
    B. 21
    C. 10
    D. 15

46. All of the following are signs of impending respiratory failure except:

    A. Accessory muscle use
    B. Respiratory rate of 18 breaths per minute
    C. Altered mental status
    D. Nasal flaring and tracheal tugging

47. Global concepts for the paramedic to consider when making an ethical decision include all of the following, except:

    A. Acknowledge the patient's autonomy
    B. Keep the patient calm
    C. Avoid harm to the patient
    D. Provide a benefit to the patient

48. _____as we know it today, stands for soundness of moral principle and character, uprightness, and honesty.

    A. Integrity
    B. Sincerity
    C. Ethics
    D. Candor

49. The development of a new type of cell with an uncontrolled growth pattern is called:

    A. Hypertrophy
    B. Metaplasia
    C. Neoplasia
    D. Hyperplasia

50. When a cell membrane ingests a substance, this is called:

    A. Endocytosis
    B. Exocytosis
    C. Phagocytosis
    D. Pinocytosis

51. Morbidly obese patients often have sleep apnea and:

    A. Multiple sclerosis
    B. Lactose intolerance
    C. Respiratory function impairment
    D. Hypolipidemia

52. The general properties of a drug include all of the following, except:

    A. Research
    B. Therapeutic
    C. Prophylactic
    D. Diagnostic

53. A drug's ability to join with a receptor is known as:

    A. Agonist
    B. Selective response
    C. Affinity
    D. Ability

54. All of the following are examples of a sharp except a (an):

    A. Used bristojet injector
    B. Open ampule
    C. Open vial
    D. Used angio catheter

55. Multiply the following fractions and reduce to the lowest terms, 8/16 x 3/10=_____.

    A. 3/20
    B. 3/5
    C. 24/40
    D. 24/160

56. If a patient received an adequate trial of ALS in the field, in which
     circumstance should you continue the arrest and transport to the
     local ED?

     A. Rigor mortis is apparent
     B. A low body temperature
     C. A lengthy downtime
     D. The patient has a mortal injury

57. If a rescuer is unable to cover both the mouth and nose of an infant
     with his/ her own mouth, it is acceptable to:

     A. Use a Combi-tube
     B. Skip the ventilations
     C. Do mouth to nose breathing
     D. Defibrillate the patient

58. What is the compression to ventilation ratio in adults?

     A. 5:1
     B. 2:15
     C. 30:2
     D. 1:5

59. Infants are primarily obligate nose breathers until what age?

     A. 6 months
     B. 4 weeks
     C. 6 weeks
     D. 3 months

60. At what age do the posterior fontanelles close on an infant?

    A. 3 months
    B. 18 months
    C. 6 months
    D. 12 months

61. A heart rate of less than _____ bpm in a newborn is abnormal.

    A. 110
    B. 120
    C. 100
    D. 140

62. What is the leading cause of death among early adults?

    A. Suicide
    B. Cancer
    C. AMI
    D. Accidents

63. The upper airway consists of all of the following structures, except:

    A. Vocal cords
    B. Cricoid ring
    C. Thyroid cartilage
    D. Alveoli

64. After a cervical collar is applied to a patient, it can be removed for intubation or to correct a problem in the airway only when:

    A. The front of a collar is removed
    B. Two hands provide manual in-line stabilization
    C. A second paramedic is available to assist
    D. The head is taped to a long backboard

65. The laryngeal mask airway, or LMA, can be used on all of the following patients, except:

    A. Elderly
    B. Chest trauma
    C. Children
    D. Cardiac arrest

66. When caring for a hearing impaired patient be sure to face the patient when you are speaking and to:

    A. Look to see if his/ her hearing aid is on
    B. Speak very slowly
    C. Use a family member to communicate
    D. Exaggerate your speech

67. While taking a blood pressure, the sounds heard with a stethoscope that indicate the systolic and diastolic readings are called:

    A. Apical pulse
    B. Korotkoff sounds
    C. Pulse pressures
    D. Pulsus paradoxus

68. Rhonchi may accompany:

    A. Asthma
    B. Emphysema
    C. Chronic bronchitis
    D. Both B and C

69. _____ release histamine that controls vasodilation/ vasoconstriction:

    A. Basophils
    B. Stem cells
    C. Eosinophils
    D. Neutrophils

70. An acute pulmonary disease that produces non-cardiogenic pulmonary edema and severe hypoxemia, and is brought about by increased capillary permeability in the pulmonary arterial system is known as:

    A. Asthma
    B. Acute pulmonary embolism
    C. Adult respiratory distress syndrome
    D. Pneumonia

71. All of the following are physical finding that are detectable by auscultation except:

    A. Heart sounds
    B. Carotid bruits
    C. Abdominal organ enlargement
    D. Bowel sounds

72. The purpose of EKG monitoring is to determine the:

    A. Presence of left ventricular hypertrophy
    B. Myocardial contractile capability
    C. Oxygen saturation in the blood
    D. Presence and type of electrical activity of the heart

73. All of the following conditions may be associated with pulseless electrical activity (PEA) except:

    A. Anginal syndrome
    B. Acute pulmonary embolism
    C. Cardiogenic shock
    D. Cardiac tamponade

74. All of the following are parts of the pre-hospital emergency medical care for a patient with an abdominal aortic aneurysm except:

    A. IV access
    B. High- concentration oxygen
    C. On-scene stabilization if the patient is hypotensive
    D. EKG monitoring

75. All of the following are examples of patients with altered mental status except:

    A. Speaking in a foreign language
    B. Narcotic overdose
    C. Seizures
    D. Syncope

76. All of the following are phases of a generalized grand mal seizure, except:

    A. Postictal phase
    B. Aura
    C. Clonic phase
    D. Hypontonic phase

77. All of the following are correct statements concerning an acute hemorrhagic stroke, except:

    A. The mortality rate is 50 – 80%
    B. Patients frequently have a history of cancer
    C. The two types are intracerebral hemorrhage and subarachnoid hemorrhage
    D. The average age is fifties to early sixties

78. All of the following are symptoms or signs of transient ischemic attacks, except:

    A. Dysuria
    B. Monocular blindness
    C. Numbness and / or paresthesias
    D. Staggering gait

79. All of the following are known causes of diabetic ketoacidosis in known diabetics, except:

    A. Taking too much insulin
    B. Infection
    C. Failing to take insulin
    D. Increased stress from surgery or trauma

80. An allergic reaction and/ or an anaphylactic reaction is usually initiated after the body is exposed to:

  A. An antibody
  B. A bacteria
  C. A medication
  D. An antigen

81. All of the following are agents that may cause anaphylaxis, except:

  A. Antibiotics, bee stings, and aspirin
  B. Nuts, eggs, and seafood
  C. Water, vasoline, and air
  D. Non-steroidal anti-inflammatory agents, aspirin, and x-ray contrast material

82. Which of the following manual airway maneuvers is the preferred method for spine injured patients?

  A. Head tilt, neck lift
  B. Modified jaw thrust
  C. Head tilt, chin lift
  D. None of the above

83. The adult dose of syrup of ipecac is:

  A. 40 ml
  B. 10 ml
  C. 20 ml
  D. 30 ml

84. The clear liquid portion of the cytoplasm is called the:

    A. Aquasol
    B. Chromosol
    C. Cytoplastic
    D. Cytosol

85. If a patient delivers her baby in the field and the baby is not progressing down the birth canal, the Paramedic should pull the baby out with force.

    A. True
    B. False

86. What is the first IV drug administered in ventricular fibrillation?

    A. Atropine
    B. Epinephrine
    C. Bretylium
    D. Lidocaine

87. The most significant clinical finding about a person's hearing is:

    A. Acute changes in hearing
    B. The use of a hearing aid
    C. The absence of hearing
    D. The presence of tinnitus

88. All of the following are considered baseline vital signs, except:

    A. Respiratory effort
    B. Blood pressure
    C. Temperature
    D. Pulse rate

89. You are evaluating a patient who has a chief complaint of respiratory distress. On which of the following body systems, besides the respiratory system, should you focus your exam?

    A. Endocrine and musculoskeletal
    B. Gastric and behavioral
    C. Cardiac and musculoskeletal
    D. Neurologic and cardiac

90. One step the paramedic can take to make a scene safe is to:

    A. Suction the patient
    B. Look for hazards
    C. Maintain C-spine stabilization
    D. Notify dispatch of arrival

91. The level of consciousness or mental status of an infant is best assessed by:

    A. Calling medical control to consult for advice
    B. How loud they cry
    C. Using the Broslow tape for pediatrics
    D. Asking the parent or caregiver to determine if the response is normal

92. Skin color is assessed by looking at the:

    A. Tongue and eyes
    B. Face and neck
    C. Nail beds, lips, and eyes
    D. Hands and ankles

93. While gathering a SAMPLE history on a medical patient, the paramedic asks about all the following, except:

    A. Pertinent blood loss
    B. Events leading up to this episode
    C. Allergies
    D. Medications

94. Common causes of vertigo include all of the following, except:

    A. Alcohol
    B. Viruses
    C. Syncope
    D. Traumatic brain injury (TBI)

95. One of the most common rooms in the home in which a patient becomes ill or injured is the:

    A. Attic
    B. Living room
    C. Bathroom
    D. Basement

96. Critical trauma patients have the best chance for survival if they can be stabilized in a surgical suite within _____ of the onset of injury.

    A. 6 hours
    B. 2 hours
    C. 1 hour
    D. 30 minutes

97. The maximum time the paramedic should spend on-scene with a critical trauma patient, barring any lengthy extrication, is _____ minutes.

    A. 30
    B. 10
    C. 20
    D. 15

98. Which of the following is the cornerstone of patient care?

    A. Scene size-up
    B. Past medical history
    C. Field impression
    D. Assessment

99. All of the following can cause a patient to be uncooperative, hindering the field assessment, except:

    A. Head injury
    B. Tunnel vision
    C. Hypoxia
    D. Hypovolemia

100. All of the following are critical life-threatening conditions the paramedic would address during an ongoing assessment, except:

    A. Major multi-system trauma
    B. Major single system trauma
    C. TIA
    D. AMI

101. Which of the following acronyms is used to help paramedics determine treatment and transportation priority decisions?

   A. PALS
   B. ALS
   C. CUPS
   D. PHTLS

102. The EMS is a professional who has been trained in each of the following, except:

   A. Triage
   B. Logistics
   C. Telephone interrogation
   D. Rapid response to calls

103. Voice transmission is also referred to as:

   A. Duplex
   B. Analog transmission
   C. Simplex
   D. Digital transmission

104. The prefix Celi/o means:

   A. Abdomen
   B. Around
   C. Neck
   D. Color

105. _____findings are information the paramedic can measure, such as vital signs:

    A. Lawful
    B. Objective
    C. Subjective
    D. Neutral

106. Two types of narrative formats used in documentation of PCR's are:

    A. SAMPLE and APGAR
    B. CUPS and AVPU
    C. SOAP and CHART
    D. PMHx and OPQRST

107. The process where oxygenated blood is pumped to the tissues, and waste products returned to the lungs, is called:

    A. Perfusion
    B. Respiration
    C. Diffusion
    D. Ventilation

108. Patients who are taking _____ typically have severe pulmonary disease.

    A. Inhaled steroids
    B. Oral corticosteroids
    C. MOA inhibitors
    D. Decongestants

109. What should be the primary concern for the paramedic when treating a patient who may have pneumonia?

    A. Respiratory distress
    B. Severe dehydration
    C. Respiratory failure
    D. Exposure to a contagious patient

110. A 25 year old female is complaining of abdominal pain that she describes as intermittent and cramping. What type of pain is the patient most likely experiencing?

    A. Neurotic
    B. Somatic
    C. Visceral
    D. Idiopathic

111. All of the following are components of the nerve cell body, except the:

    A. Myelin
    B. Soma
    C. Dendrite
    D. Axon

112. The endocrine system is an integrated_____and coordination system enabling reproduction, growth and development, and regulation of energy:

    A. Nerve
    B. Chemical
    C. Fluid
    D. Muscle

113. Diabetic patients do not always have the classic symptoms of myocardial ischemia such as crushing substernal chest pain because:

    A. Elevated blood lipid levels alter sensation
    B. Many diabetics have some form of neuropathy
    C. Glucose is the sole source of oxidative metabolism for the CNS
    D. Insulin numbs the pain

114. The onset of action is faster and duration is shorter with_____insulin preparations:

    A. Chicken
    B. Beef
    C. Pork
    D. Human

115. Which of the vital signs findings is not usually associated with an allergic reaction?

    A. Hypotension
    B. Tachycardia
    C. Tachypnea
    D. Bradycardia

116. A spasmodic contraction of the diaphragm is a:

    A. Stridor
    B. Cough
    C. Hiccup
    D. Sneeze

117. Which one of the following is true regarding psychiatric emergencies?

   A. More females succeed at suicide
   B. The supine position is the best method of transporting a violent, restrained patient.
   C. In managing a violent patient, it is advisable to wait for assistance
   D. Individuals under the influence of medications, drugs and/ or alcohol have predictable behavior problems.

118. What does treatment for a patient who is experiencing stable angina consist of?

   A. Seat patient upright, start an IV, EKG, and lasix
   B. Rest, oxygen, nitroglycerin administration
   C. Reassurance, oxygen, EKG monitoring, and morphine sulfate
   D. Oxygen and defibrillation if the ORS complex is wide

119. Which of the following situations represents "expressed" consent?

   A. The patient says "I don't need any help. Just let me die."
   B. The patient says "Help me. My chest hurts."
   C. The patient is unconscious and unresponsive
   D. The patient is 10 years old with no guardian present

120. Phenobarbital is an example of what class or type of medication?

   A. Beta agonist agent
   B. Hypoglycemic
   C. Sedative or anticonvulsant
   D. Synthetic form of insulin

121.   In the case of a cardiac arrest patient, epinephrine is used to:

      A.   Increase ventricular fibrillation amplitude
      B.   Increase the effects of the other drugs
      C.   Prevent anaphylaxis from lidocaine reactions
      D.   Increase the blood pressure

122.   A patient's signs and symptoms include orthopnea, agitation, spasmodic coughing, jugular vein distention, rales, cyanosis, and elevated pulse, blood pressure, and respirations. What condition should you suspect?

      A.   Cardiogenic shock
      B.   Left sided heart failure
      C.   Myocardial infarction
      D.   Right sided heart failure

123.   What is sodium nitroprusside (Nipride) used in the treatment of?

      A.   Cardiogenic shock
      B.   Deep venous thrombosis
      C.   Myocardial infarction
      D.   Hypertensive emergency

124.   All of the following are vasopressors except:

      A.   Atropine
      B.   Norepinephrine
      C.   Aramine
      D.   Dopamine

125.   The most serious complication of a prolapsed cord is:

        A.   Laceration of the cord
        B.   Compromised blood flow to infant
        C.   Hemorrhagic shock to the mother
        D.   Hemorrhagic shock to the infant

ANSWER SHEET 6

1.   B- NSAIDS
2.   D- Expired air
3.   A- Approximately 40% of near-drowning victims do not aspirate water
4.   B- The second prominent structure felt
5.   C- Edema of the glottis
6.   A- Stem cells
7.   D- Exhilaration
8.   B- Crying
9.   A- High abuse potential and no medical uses
10. B- Right
11. B- Low
12. A- Miller
13. C- What type of pain ore you experiencing?
14. A- Brachial
15. D- 11
16. C- T4
17. B- Elbow
18. C- Spleen
19. B- HIV
20. B- Catabolism
21. D- 60%
22. A- Acid
23. B- Diaphoresis
24. C- COPD
25. B- Stenotic
26. D- Paradoxus
27. A- Ischemia
28. B- V1 – V2
29. A- Dysrhythmias
30. C- Synapses

31. C- Accommodation
32. A- Late
33. C- Death
34. D- Pancreas
35. A- Infection
36. B- Histamine
37. C- Somatic
38. D- Esophageal varices
39. B- Cocaine
40. A- Absorption
41. D- Activated charcoal
42. C- Metabolic
43. D- A and B only
44. D- All of the above
45. B- 21
46. B- Respiratory rate of 18 breaths per minute
47. B- Keep the patient calm
48. A- Integrity
49. C- Neoplasia
50. A- Endocytosis

ANSWER SHEET 6

51. C- Respiratory function impairment
52. A- Research
53. C- Affinity
54. C- Open vial
55. A- 3/20
56. B- A low body temperature
57. C- Do mouth to nose breathing
58. C- 30:2
59. B- 4 weeks
60. A- 3 months
61. C- 100
62. D- Accidents
63. D- Alveoli
64. B- Two hands provide manual in-line stabilization
65. B- Chest trauma
66. A- Look to see if his/ her hearing aid is on
67. B- Korotkoff sounds
68. D- Both B and C
69. A- Basophils
70. C- Adult respiratory distress syndrome
71. C- Abdominal organ enlargement
72. A- Presence of left ventricular hypertrophy
73. A- Anginal syndrome
74. C- On-scene stabilization if the patient is hypotensive
75. A- Speaking in a foreign language
76. D- Hypotonic phase
77. B- Patients frequently have a history of cancer
78. A- Dysuria
79. A- Taking too much insulin
80. D- An antigen
81. C- Water, vasoline, and air

82. B- Modified jaw thrust
83. D- 30 ml
84. D- Cytosol
85. B- False
86. B- Epinephrine
87. A- Acute changes in hearing
88. C- Temperature
89. C- Cardiac and musculoskeletal
90. B- Look for hazards
91. D- Asking the parent or caregiver to determine if the response is normal
92. C- Nail beds, lips and eyes
93. A- Pertinent blood loss
94. C- Syncope
95. C- Bathroom
96. C- 1 hour
97. B- 10
98. D- Assessment
99. B- Tunnel vision
100. C- TIA

ANSWER SHEET 6

101. C- CUPS
102. C- Telephone interrogation
103. B- Analog transmission
104. A- Abdomen
105. B- Objective
106. C- SOAP and CHART
107. A- Perfusion
108. B- Oral corticosteroids
109. A- Respiratory distress
110. C- Visceral
111. A- Myelin
112. B- Chemical
113. B- Many diabetics have some form of neuropathy
114. D- Human
115. C- Tachypnea
116. C- Hiccup
117. C- In managing a violent patient, it is advisable to wait for assistance
118. B- Rest, oxygen, nitroglycerin administration
119. B- The patient says "Help me. My chest hurts."
120. C- Sedative or anticonvulsant
121. B- Increase ventricular fibrillation amplitude
122. B- Left sided heart failure
123. D- Hypertensive emergency
124. A- Atropine
125. B- Compromised blood flow to infant

Test 7

1. The patient who subconsciously converts his anxiety into a body dysfunction suffers from:

    A. Conversion hysteria
    B. Incontinence
    C. Lack of control
    D. Tendency to cry easily

2. The chemical mediator for the parasympathetic nervous system is:

    A. Vagus
    B. Propranolol
    C. Norepinephrine
    D. Acetylcholine

3. A harsh upper airway sound that can be heard when the patient inhales is called:

    A. Dyspnea
    B. A cough
    C. Stridor
    D. Dyphonia

4. Listening skills that a paramedic should hone with frequent practice, include all of the following, Except:

    A. Ambient noises
    B. Lung sounds
    C. Bruits
    D. Heart sounds

5. Bleeding described as "spurting bright red" is usually from a/ an:

    A. Artery
    B. Vesicle
    C. Vein
    D. Capillary

6. Two common examples of soft tissue trauma that ay be fatal are hematomas and:

    A. Insect bites
    B. Secondary infections
    C. Abrasions
    D. Genital warts

7. The three classifications of burn severity do not include:

    A. Severe
    B. Minor
    C. Moderate
    D. Eschar

8. Which of the following is not a common MOI for blunt throat injuries?

    A. Stabbing
    B. Hanging
    C. MVC
    D. Strangulation

9. The middle ear is separated from the external canal by the:

    A. Eardrum
    B. Pinna
    C. Inner ear
    D. Cartilage

10. The second cervical vertebrae is also called the:

    A. Pivotal
    B. Cribiform
    C. Axis
    D. Atlas

11. Sensation in the little finger is located in which nerve root?

    A. L-1
    B. C-3
    C. T-4
    D. C-8

12. All of the following are major arteries of the chest, except:

    A. Inferior vena cava
    B. Aorta
    C. Carotid artery
    D. Internal mammary artery

13.   Which of the following are both semilunar valves?

      A.   Aortic and pulmonic
      B.   Tricuspid and pulmonic
      C.   Aortic and bicuspid
      D.   Biscuspid and pulmonic

14.   The primary symptom of a fracture is:

      A.   Paresthesia
      B.   Pain
      C.   Deformity
      D.   Discoloration

15.   All of the following are signs of shock, except:

      A.   Restlessness
      B.   Cold, clammy skin
      C.   Constriction of pupils
      D.   Ashen or cyanotic skin coloration

16.   When examining the abdomen:

      A.   Check for function masses
      B.   Palpate before inspecting
      C.   Check for Battle's sign
      D.   Inspect before palpation

17. Administration of a 1 to 2 gram IV dose of Magnesium Sulfate is recommended for:

    A. Torsade de pointes
    B. PEA
    C. Ventricular fibrillation
    D. Ventricular tachycardia

18. Which of the following would be considered a parasympathetic blocker?

    A. Epinephrine
    B. Inderal
    C. Isuprel
    D. Atropine

19. The Paramedic must hold the needle at a _____ degree angle just under the skin for the subcutaneous injection:

    A. 45 degree
    B. 25 degree
    C. 15 degree
    D. 35 degree

20. In a patient with severe hypothermia, movement or jostling may trigger which of the following?

    A. Respiratory arrest
    B. Vomiting and aspiration
    C. Ventricular fibrillation
    D. Seizures

21. Chris has had excessive vomiting for five days, you would expect:

    A. Acidocis
    B. Normal Ph
    C. Alkalosis
    D. All of the above

22. The pituitary gland is located:

    A. At the base of the brain
    B. Below each kidney
    C. In the center of the brain
    D. Lateral and anterior to the trachea

23. The breakdown of glucose in the liver is known as:

    A. Glucogenolysis
    B. Glucogenseis
    C. Gluconeogenesis
    D. Glucolysis

24. Which of the following is the best definition of frostbite?

    A. A bite like injury occurring in frosty snow
    B. A rare insect bite occurring in frosty snow
    C. A localized injury due to freezing of body tissues
    D. Hypothermia induced by prolonged exposure to frosty snow

25. Which of the following biological agents has the capability of being transmitted person to person?

     A. Smallpox
     B. Anthrax
     C. Brucellosis
     D. Tularemia

26. How should you control bleeding after the normal delivery of an infant?

     A. Perform fundal massage
     B. Elevate the pelvis
     C. Pack the vagina with sterile gauze
     D. Apply direct pressure to the genitalia

27. Which of the following represents a significant mechanism of injury?

     A. A child falls from a 7 foot high ledge
     B. A small child is in a 45 mph car accident
     C. An adult pedestrian is hit by a bicycle
     D. An adult falls from a 6 foot high ledge

28. The first step in triage at an MCI is to:

     A. Evaluate the victim's mental status and ABC's
     B. Assess the victim's hemodynamic status and AVPU
     C. Assess the victim's respiratory status and pulse rate
     D. Direct the walking wounded away from the scene

29. Patients who have overdosed and have altered mental status are best transported:

   A. Full Fowler's position
   B. Prone position
   C. Supine position
   D. Left lateral recumbent position

30. Your patient is a 45 year old man who has been in an automobile crash. His respiratory rate is 34 with normal chest wall expansion: Systolic blood pressure is 78, capillary refill is delayed, and his Glasgow Coma Scale score is 10. This patient's revised Trauma Score is:

   A. 10
   B. 5
   C. 9
   D. 4

31. What is a positive Battle's sign an indication of?

   A. Basilar skull fracture
   B. Cervical spinal trauma
   C. Orbital skull fracture
   D. Subarachnoid hemorrhage

32. Which dysrhythmia is considered normal in an athletic adult?

   A. First degree AV block
   B. Sinus bradycardia
   C. Sinus arrest
   D. Atrial flutter

33. What is the primary goal of management for a patient with left ventricle failure and pulmonary edema?

    A. Prevent serious dysrhythmias
    B. Decreased cardiac output
    C. Decreased venous return
    D. Initiate thrombolytic therapy

34. What is the usual dosage and route of administration of diazepam given before synchronized cardioversion?

    A. 5 – 15 mg by slow IV push
    B. 2 – 5 mg given intramuscularly
    C. 10 – 15 administered rectally
    D. 5 – 10 mg directly into the vein

35. What is the role of beta agonists like albuterol in the treatment of anaphylaxis?

    A. To reverse bronchospasm
    B. To relieve anxiety
    C. To raise blood pressure
    D. To prevent shock

36. A whistling or musical sound heard on exhalation is referred to as what abnormal breath sound?

    A. Stridor
    B. Wheezing
    C. Friction rub
    D. Snoring

37.   Rapid cooling for a heat stroke patient is to prevent:

    A.   Seizures
    B.   Heart failure
    C.   Irreversible brain damage
    D.   Myocardial Infarction

38.   Narcan is used to counteract the effects of:

    A.   Cocaine and heroin
    B.   Alcohol and codeine
    C.   Darvon and morphine
    D.   Morphine and elavil

39.   A victim presenting with left sided chest pain associated with shortness of breath, is found to be cyanotic. Breath sounds are more distant over the left lung than over the right. These findings might indicate:

    A.   Spontaneous pneumothorax
    B.   Pulmonary edema
    C.   Acute Myocardial infarction
    D.   Pneumonia

40.   The best way to determine if the heart is pumping is to:

    A.   Check the blood pressure
    B.   Check the pulse
    C.   Check the pupils
    D.   Check the EKG

41. Your patient is 11 months old. She has wheezing and tachypnea and a fever of 100.8 F. What do you suspect is wrong with the patient?

    A. Croup
    B. Asthma
    C. Epiglottitis
    D. Bronchiolitis

42. Cardiac arrest in young children is most commonly associated with which of the following?

    A. Trauma from automobile accidents
    B. Respiratory problems or diseases
    C. Underlying cardiac disease processes
    D. Burn trauma from building fires

43. The _____ is the largest organ in the body and often sustains injuries during rapid deceleration forces.

    A. Lung
    B. Liver
    C. Kidney
    D. Heart

44. The scapula is broken down into the following areas, except for the:

    A. Upper division
    B. Glenoid fassa
    C. Lower division
    D. Median division

45. When muscles contract, they pull the _____, which then causes the bones to move at the joints.

    A. Ligaments
    B. Tendons
    C. Haversian canals
    D. Marrow

46. There are three causes of infant apnea: obstructive, central and:

    A. Incomplete
    B. Mixed
    C. Anomaly
    D. Complete

47. Which of the following is not a risk factor associated with crack/cocaine use by a pregnant woman?

    A. Placenta previa
    B. Abruptio placentia
    C. Miscarriage
    D. Premature labor

48. During the delivery of twins, the paramedic can tell if the babies are fraternal or identical because fraternal twins:

    A. Develop from the same zygote
    B. Have their own umbilical cords
    C. Always have their own placenta
    D. (One) will present as a breach

49.   Complications from vomiting in the neonate include:

      A. Electrolyte imbalance
      B. Seizures
      C. Increased ICP
      D. Lactose intolerance

50.   Which of the following groups are living longer?

      A. People who have attained higher educations
      B. Married people
      C. Cohabitating people
      D. All of the above

51.   Which of the following conditions may result in a transient blindness
      for the patient?

      A. TIA
      B. Acute hypothermia
      C. Acute head injury
      D. Near drowning

52.   When managing patients with Down syndrome, the paramedic
      should keep in mind all of the following, except:

      A. They have abnormal intestines
      B. These patients have a below average IQ
      C. They have a unique airway anatomy
      D. That many have a congenital heart defect

53. Assessment and care of a patient who is a victim of sexual assault
     should include which of the following?

     A. Allow the patient to bathe and douche
     B. Perform a complete vaginal exam
     C. Ask detailed questions about the assault
     D. Place sterile dressings on any wounds

54. It is estimated that one in _____ Americans meets the definition
     for obesity.

     A. Three
     B. Two
     C. Five
     D. Seven

55. Obesity is defined as being _____ above ideal body weight.

     A. 30 – 40%
     B. 7 – 12%
     C. 20 – 30%
     D. 10 – 20%

56. _____standards are minimum standards and not considered
     the "gold standard."

     A. Regional
     B. State
     C. National
     D. Local

57. One concept that appears in most laws in statutes that deal with emergency vehicle operation is the concept of:

     A. Causation
     B. Res ipsa locquitor
     C. Negligence
     D. Due regard

58. When your ambulance is the first to arrive at the scene of a highway incident and no potential hazards are apparent, you should park your emergency vehicle at least _____ the accident;

     A. 100 feet behing
     B. 50 feet in front of
     C. 50 feet behind
     D. 100 feet in front of

59. Eye protection is best provided by:

     A. A fire helmet face shield
     B. Contact lenses
     C. Industrial safety glasses
     D. Regular glasses

60. Upon arrival at the rescue incident, the EMS crew should:

     A. Conduct a scene size up
     B. Set out flares on the roadway
     C. Determine the hospital destination
     D. Disentangle the patient

61. Alcohol is a contributory factor to as many as _____% of boating fatalities.

    A. 75
    B. 15
    C. 35
    D. 50

62. The first standard to guide hazardous material operations specifically for EMS providers was NFPA:

    A. 1500
    B. 473
    C. 4 72
    D. 704

63. Water is a universal decon solution that dilutes and reduces _____ absorption.

    A. Topical
    B. Parenteral
    C. Enteral
    D. Metabolic

64. Which of the following devices provides a continuous visual display of the level of expired CO2?

    A. Colorimetric device
    B. Capnograph
    C. Capnometer
    D. All of the above

65.   In caring for a burn patient who is in need of fluid resuscitation, you
       should select which of the solutions listed below?

       A. D50W
       B. ½ normal saline
       C. D5W
       D. Normal saline

66.   The normal arterial PO2 (at sea level) is:

       A. 80 – 100 mmHg (torr)
       B. 8.0 – 10.0 mmHg (torr)
       C. 7.35 – 7.40 mmHg (torr)
       D. 35 – 40 mmHg (torr)

67.   The normal resting respiratory rate for an adult is ____respiration's
       per minute.

       A. 17 – 25
       B. 6 – 10
       C. 12 – 20
       D. 13

68.   An exaggerated immune response to an environmental allergen is
       called:

       A. Isoimmunity
       B. Autoimmunity
       C. Allergy
       D. All of the above

69. The initial management of a comatose patient includes:

    A. Consider thiamine administration
    B. Give bolus of Dextrose 50%
    C. Give Narcan 0.4 mg IV bolus
    D. Assure respiratory function, assisting as needed

70. Patients who have been trapped in an enclosed space with a combustion will most likely suffer from:

    A. Heat stroke
    B. Cyanide poisoning
    C. Carbon monoxide poisoning
    D. Thermal burns

71. All of the following are components of the past medical history except:

    A. Patient allergies
    B. Onset of pain
    C. Medications taken
    D. Events preceding the illness or injury

72. You must administer 1000 ml of 0.9% sodium chloride to a patient over a period of 6 hours, using a 1000 ml bag of solution and a 10 gtt/min drip set. How many drops per minute should you administer to deliver this amount:

    A. 28 gtt/min
    B. 44 gtt/min
    C. 30 gtt/min
    D. 44 gtt/min

73. The most common EKG finding in angina pectoris is:

    A. ST- segment depression
    B. Q waves
    C. PR- interval prolongation
    D. Tall, peaked T- waves

74. Which of the following is the correct definition of an aneurysm?

    A. Increased curvature of an artery
    B. Blockage of an artery
    C. Weakening and dilation of a vessel wall
    D. Congenital defect of a vein

75. Neurologic emergencies are often a consequence of all of the following except:

    A. Intravascular protein concentration changes
    B. Head trauma
    C. Intracranial pressure changes
    D. Circulatory changes

76. All of the following are correct statement concerning type II diabetes mellitus except:

    A. It usually occurs in obese patients
    B. Obesity causes an increase in blood sugar by converting protein into sugar
    C. Most patients are treated with diet and oral agents
    D. It usually occurs in later life, not childhood

77. Hypoadrenalism, also known as Addison's disease, may present with all of the following signs and symptoms except:

    A. Weakness and fatigue
    B. Hypoglycemia
    C. Nausea and vomiting
    D. Hypertension

78. All of the following are signs or symptoms of hypothyroidism except:

    A. Dry skin and brittle hair
    B. Lethargy
    C. Cold intolerance
    D. Altered mental status

79. All of the following are routes of exposure for the potential transmission of infectious diseases except:

    A. Fecal- oral
    B. Airborne
    C. Bloodborne
    D. Talking to an exposed patient

80. All of the following are possible complications and emergencies associated with patients who have tracheostomies exept:

    A. Sepsis
    B. Tracheal stenosis
    C. Skin rash on all extremities
    D. Respiratory distress

81. An unconscious patient who has one dilated pupil that is reactive to light is showing early signs of:

    A. Status epilepticus
    B. Transient ischemic attacks
    C. Cerebral artery aneurysm
    D. Increased intracranial pressure

82. The pain of stable angina is brought on by:

    A. Overuse of nitroglycerin
    B. Exercise or stress
    C. Difficulty breathing
    D. Imminent acute myocardial infarction

83. A contrecoup contusion is a brain injury that :

    A. Results from open skull fracture and brain bruising
    B. Is on the opposite side of the head from the impact site
    C. Results from cerebral edema at the site of impact
    D. Causes subdural or epidural hematoma formation

84. Vagal maneuvers are used to treat which type of dysrhythmias?

    A. Atrial fibrillation of flutter
    B. Ventricular tachycardia
    C. Paroxysmal supraventricular tachycardia
    D. Premature atrial contractions

85.　A late sign of hypoxia in children is:

    A.　Bradycardia
    B.　Tachypnea
    C.　Tachycardia
    D.　Hypotension

86.　What is the typical presentation of a patient with esophageal varices?

    A.　Pain radiating to the jaw
    B.　Painless hematemesis
    C.　Melena and hemoptysis
    D.　Acute abdominal pain

87.　Carotid sinus massage is used for patient with which dysrhythmia?

    A.　Second degree AV block (Mobitz II)
    B.　Non-perfusing ventricular tachycardia
    C.　Paroxysmal supraventricular tachycardia
    D.　Refractory ventricular fibrillation

88.　After the first round of ACLS drugs, the patient converts to a sinus bradycardia at a rate of 40 beats per minute. What is the next drug of choice for managing this patient's bradycardia?

    A.　Dobutamine
    B.　Atropine
    C.　Isoproterenol
    D.　Epinephrine

89. Hyperventilation syndrome most often occurs in a patient who is:

    A. Asthmatic
    B. A heavy smoker
    C. Anxious and upset
    D. In shock from trauma

90. The coccyx consists of _____ vertebrae that are fused together:

    A. 12
    B. 5
    C. 3 to 5
    D. 7

91. The initial assessment includes:

    A. Responsiveness, airway, breathing, circulation
    B. Airway, breathing, circulation and hemorrhage control
    C. Airway, breathing, circulation and secondary exam
    D. Airway, breathing, circulation and treatment

92. The collection of blood in front of a patient's pupil and iris due to direct trauma is called:

    A. Hyphema
    B. Anterior chamber hematoma
    C. Retinal detachment
    D. Icterus

93. The mechanism of injury in the blast which causes injury is:

    A. Flying debris
    B. A pressure wave
    C. The patient being thrown into objects
    D. All of the above

94. A woman who is pregnant for the first time is referred to as:

    A. Primapara
    B. Primagravida
    C. Multipara
    D. Multigravida

95. An initial Alpha effect of a drug would be:

    A. Increased force of cardiac contraction
    B. Vasoconstriction
    C. Vasodilation
    D. Increased rate of cardiac contraction

96. _____ has been added to the management of potentially fatal pediatric dysrhythmias:

    A. Vasopressin
    B. Verapamil
    C. Amiodarone
    D. Cardizem

97.   Cocaine overdose has been shown to be associated with serious:

     A. Hypothermia
     B. Ventricular dysrhythmias
     C. Atrial dysrhythmias
     D. Hypomagnesium

98.   A patient has been burned on the entire anterior trunk of the body, both arms (front and back) and the face and anterior neck: what % of burn is it?

     A. 50%
     B. 18%
     C. 25%
     D. 40.5%

99.   What communications system has the capability to send and receive voice and telemetry simultaneously?

     A. Duplex
     B. Multiplex
     C. VHF
     D. UHF

100.   Continual re-experiencing of a traumatic event is a characteristic of which of the following?

     A. Stress and burnout
     B. An anxiety disorder
     C. Delayed stress reaction
     D. Cumulative stress reaction

101. To adequately ventilate a patient with a partial laryngectomy through a stoma, you should:

    A. Use a special bag valve mask designed to ventilate a stoma
    B. Pinch the nose and close the mouth
    C. Suction the stoma with a soft tip suction catheter first
    D. Use more pressure to produce adequate chest rise

102. The initial joules dosage for pediatric defibrillation is:

    A. 4 joule/kg
    B. 1 joule/kg
    C. 2 joule/kg
    D. 3 joule/kg

103. Patients who have overdosed and have altered mental status are best transported:

    A. Left lateral recumbent
    B. Supine
    C. Prone
    D. Full Fowler's

104. Which sector officer will coordinate with police to block streets and provide access at an MCI?

    A. Supply officer
    B. Triage officer
    C. Transportation officer
    D. Staging officer

105. Which statement characterizes normal physiologic changes that occur in vital signs during pregnancy?

    A. Blood pressure falls, pulse rate rises
    B. Blood pressure rises, pulse rate falls
    C. Blood pressure falls, pulse rate falls
    D. Blood pressure rises, pulse rate rises

106. An individual with known drug abuse is in respiratory distress, hypotensive, and stuporous. He has pinpoint pupils. The paramedic might suspect:

    A. Amphetamine overdose
    B. Morphine overdose
    C. Elavil overdose
    D. Valium overdose

107. The Florida Emergency Medical Services Act, providing legislation governing all pre-hospital emergency medical services is known as:

    A. Florida Statutes Chapter 401
    B. Florida Statutes Chapter 301
    C. Florida Statutes Chapter 10D-66
    D. Florida Statures Chapter 93-154

108. Spleen injury is characterized by which of the following?

    A. Left middle chest pain
    B. Right shoulder pain
    C. Left upper quadrant pain
    D. Right upper quadrant pain

109. Patients who are found in a hazardous materials incident should be initially treated in which containment zone?

    A. Moderate
    B. Hot
    C. Warm
    D. Cold

110. Narcan is given to patients who are suspected of having which of the following conditions?

    A. Narcotic overdose
    B. Increased intracranial pressure
    C. Wernicke's syndrome
    D. Korsakoff's psychosis

111. How do you break the rear window of a car?

    A. Center punch to the corner
    B. Break with an ax
    C. Hurst tool
    D. Center punch to middle of the window

112. If a paramedic elects to help an injured person and then leaves them before other help is available, she may be sued by the patient for:

    A. Non-feasance
    B. Abandonment
    C. Malfeasance
    D. Dereliction of duty

113. In a pediatric arrest, the recommended initial dose of sodium bicarbonate is:

      A. 0.5 ml of 1:10,000 solution
      B. 1 mg/kg of body weight
      C. 1 mEq/kg of body weight
      D. 0.1 mEq/kg of body weight

114. What do you do if during transmission you lose radio contact with the hospital?

      A. Switch to a different channel
      B. Request another ambulance
      C. Pull off the road and attempt to locate the problem
      D. Continue to transport and follow standing protocols

115. Of the following, which is not required for a successful legal action of negligence:

      A. Intent
      B. Harm
      C. Proximate causation
      D. Breech of duty

116. A child has overdosed on Aspirin. You would expect all of the following except:

      A. Diaphoresis and fever
      B. Hypoventilation
      C. Metabolic acidosis
      D. Vomiting and dehydration

117. Which of the following do no have "P" waves?

     A. 1st degree AV block
     B. Sinus tachycardia
     C. Atrial fibrillation
     D. Premature atrial contractions

118. One pint or unit of blood equals to how many ml's

     A. 1000 ml
     B. 250 ml
     C. 500 ml
     D. 100 ml

119. Which is not part of a radio system?

     A. Repeater
     B. Mobile transmitter
     C. Telemetry
     D. Graphic equalizer

120. Laryngeo-tracheo-bronchitis is another name for:

     A. Croup
     B. Tonsillitis
     C. Bronchiolitis
     D. Epiglottitis

121. The delivery of the placenta marks the end of the:

     A. Fourth stage of labor
     B. First stage of labor
     C. Second stage labor
     D. Third stage of labor

122.　Drug dosages are lower in elderly patients than in young adult primarily because elderly patients:

　　　A.　Have a slower rate of elimination of drugs
　　　B.　Weigh less on average than younger patients
　　　C.　Do not respond to drugs as well as the young
　　　D.　Forget they took their medication and overdose

123.　The pharyngo-tracheal lumen airway should be removed if the patient:

　　　A.　Has poor compliance
　　　B.　Vomits
　　　C.　Becomes tachycardic
　　　D.　Regains consciousness

124.　Focused examination of the abdomen of a patient who is complaining of abdominal pain should consist of:

　　　A.　Auscultation of the area of discomfort
　　　B.　Gentle palpation of the entire abdomen
　　　C.　Repeated tests for rebound tenderness
　　　D.　Percussion on the entire abdomen

125.　The best way to break a front windshield is:

　　　A.　Fire axe
　　　B.　Center punch to the corner of the windshield
　　　C.　Center punch to the middle of the windshield
　　　D.　Hurst tool

ANSWER SHEET 7

1.   A- Conversion hysteria
2.   D- Acetylcholine
3.   C- Stridor
4.   A- Ambient noises
5.   A- Artery
6.   B- Secondary infections
7.   D- Eschar
8.   A- Stabbing
9.   A- Eardrum
10.   C- Axis
11.   D- C-8
12.   A- Inferior vena cava
13.   A- Aortic and pulmonic
14.   B- Pain
15.   C- Constriction of pupils
16.   D- Inspect before palpation
17.   A- Torsades de pointes
18.   D- Atropine
19.   A- 45 degrees
20.   C- Ventricular fibrillation
21.   C- Alkalosis
22.   A- At the base of the brain
23.   A- Glucogenolysis
24.   C- A localized injury due to freezing of body tissues
25.   A- Smallpox
26.   A- Perform fundal massage
27.   B- A small child is in a 45 mph car accident
28.   D- Direct the walking wounded away from the scene
29.   D- Left lateral recumbent position
30.   C- 9
31.   A- Basilar skull fracture

32. B- Sinus bradycardia
33. C- Decreased venous return
34. A- 5 – 15 mg by slow IV push
35. A- To reverse bronchospasm
36. B- Wheezing
37. B- Heart failure
38. C- Darvon and morphine
39. A- Spontaneous pneumothorax
40. B- Check the pulse
41. D- Bronchiolitis
42. B- Respiratory problems or diseases
43. B- Liver
44. D- Median division
45. B- Tendons
46. B- Mixed
47. A- Placenta previa
48. C- Always have their own placenta
49. A- Electrolyte imbalance
50. D- All of the above

ANSWER SHEET 7

51. C- Acute head injury
52. A- They have abnormal intestines
53. D- Place sterile dressing on any wounds
54. A- Three
55. C- 20 – 30%
56. B- State
57. D- Due regard
58. B- 50 feet in front of
59. C- Industrial safety glasses
60. A- Conduct a scene size up
61. D- 50
62. B- 473
63. A- Topical
64. B- Capnograph
65. D- Normal saline
66. A- 80 – 100 mmHg (torr)
67. C- 12 – 20
68. C- Allergy
69. D- Assure respiratory function, assisting as needed
70. C- Carbon monoxide poisoning
71. B- Onset of pain
72. A- 28 gtt/min
73. A- ST- segment depression
74. C- Weakening and dilation of a vessel wall
75. A- Intravascular protein concentration changes
76. B- Obesity causes and increase in blood sugar by converting protein into sugar
77. D- Hypertension
78. A- Dry skin and brittle hair
79. D- Talking to an exposed patient
80. C- Skin rash on all extremities

81. D- Increased intracranial pressure
82. B- Exercise or stress
83. B- Is in the opposite side of the head from the impact site
84. C- Paroxysmal supraventricular tachycardia
85. A- Bradycardia
86. B- Painless hematemesis
87. C- Paroxysmal supraventricular tachycardia
88. D- Epinephrine
89. C- Anxious and upset
90. C- 3 to 5
91. B- Airway, breathing, circulation and hemorrhage control
92. A- Hyphema
93. D- All of the above
94. B- Primagravida
95. B- Vasoconstriction
96. C- Amiodarone
97. B- Ventricular hysrhythmias
98. D- 40.5%
99. B- Multiplex
100. C- Delayed stress reaction

ANSWER SHEET 7

101. B- Pinch the nose and close the mouth
102. C- 2 joule/kg
103. A- Left lateral recumbent
104. D- Staging officer
105. A- Blood pressure falls, pulse rate rises
106. B- Morphine overdose
107. A- Florida Statutes Chapter 401
108. D- Right upper quadrant pain
109. C- Warm
110. A- Narcotic overdose
111. A- Center punch to the center
112. B- Abandonment
113. C- 1 mEq/kg of body weight
114. D- Continue to transport and follow standing protocols
115. A- Intent
116. B- Hypoventilation
117. C- Atrial fibrillation
118. C- 500 ml
119. D- Graphic equalizer
120. A- Croup
121. D- Third stage of labor
122. A- Have a slower rate of elimination of drugs
123. D- Regains consciousness
124. B- Gentle palpation of the entire abdomen
125. A- Fire axe

Test 8

1. A coral snake can be recognized by its color which are:

    A. Red, orange, black
    B. Yellow, red, white
    C. Black, yellow, red
    D. Black, red, white

2. A stroke victim should be transported:

    A. In a position of comfort
    B. Affected side up
    C. Affected side down
    D. On their back

3. When encountering a patient who is disoriented, the paramedic should:

    A. Focus on transporting the patient to the nearest psychiatric facilty
    B. Attempt to keep the patient aware of the time, place, person and situation
    C. Ignore the patient's disorientation and proceed to care for the patient's physical needs
    D. Restrain the patient to prevent problems

4. Treatment of abdominal evisceration should include all but which one of the following?

    A. IV of Ringer's lactate
    B. Replace and cover eviscerating organs
    C. Treat for shock, including the use of Oxygen
    D. Keep the organs moist and stable with sterile sponges

5. Clinical signs of tension pneumothorax are:

    A. Increased physical movement of the chest on the injured side
    B. Dyspnea
    C. Mediastinal shift
    D. Both B and C

6. A patient is covered with alpha-radioactive material after an accidental spill. An adequate level of shielding would be:

    A. Aluminum foil
    B. A lead apron
    C. A cloth uniform
    D. A concrete wall

7. What is the correct field treatment for a frostbitten body part?

    A. Cover the frozen part tightly win wet occlusive dressings
    B. Transport the patient to the hospital
    C. Rub the affected part with crushed ice or snow until warmed
    D. Warm the affected part in water maintained at 100 – 106 degrees

8. Which of the following can complicate ventilation in a pediatric patient?

    A. Hyperextension of the neck
    B. Ventilatory pressure that is higher than used for adults
    C. Cricoid pressure
    D. A bag valve mask without a pop off valve

9. The main difference between the psychotic and neurotic patient is:

    A. Neurotic patient is not in touch with reality
    B. The psychotic patient is always more dangerous
    C. The psychotic patient is not in touch with reality
    D. The psychotic patient has hallucinations

10. A harsh, high pitched respiratory sound associated with severe upper airway obstruction such as laryngeal edema is called:

    A. Croup
    B. Rales
    C. Wheezing
    D. Stridor

11. How should you control bleeding after the normal delivery of an infant?

    A. Pack the vagina with sterile gauze
    B. Perform fundal massage
    C. Apply direct pressure to the genitalia
    D. Elevate the pelvis

12.   Black, tarry stools are an indication of:

     A.   Appendicitis
     B.   GI hemorrhage
     C.   Hemorrhoids
     D.   Cholecystitis

13.   What does a carotid artery bruit indicate?

     A.   Obstruction of blood flow
     B.   Congestive heart failure
     C.   Good peripheral perfusion
     D.   Jugular vein distention

14.   You have arrived at the scene of a building collapse involving at least
     40 victims. A patient is in cardiac arrest. As the first paramedic on
     the scene you should classify this patient as:

     A.   Second priority
     B.   Highest priority
     C.   Lowest priority
     D.   Delayed priority

15.   Which of the following is the most appropriate initial joules setting
     for defibrillating pulse-less ventricular tachycardia?

     A.   200 joules
     B.   300 joules
     C.   50 joules
     D.   400 joules

16. Which statement about the use of Nitronox in the field is correct?

     A. Nitronox may be safely used for patients with head injury
     B. Nitronox is a short acting agent that is administered via
         inhalation
     C. Nitronox is used to manage pain in patients with COPD
         and asthma
     D. Nitronox may be given to patients with acute abdominal
         distension

17. All of the following are signs or symptoms of a pulmonary embolus
     except:

     A. Tachycardia and tachypnea
     B. Hypertension
     C. Sudden onset of unexplained dyspnea
     D. Sharp chest pain

18. All of the following could reduce radio transmission capabilities
     except:

     A. Weak radio battery
     B. No repeater
     C. Hold the antenna in a horizontal position
     D. Holding the radio in a vertical position

19. The digital intubation method is used for patients who:

     A. Have arthritis is the neck
     B. Have suspected spinal injury
     C. Have short anterior cords
     D. Are very old or very young

20. The first step in immobilizing a patient on a short spine board is to:

    A. Apply manual stabilization
    B. Place board behind patient
    C. Remove patient from the car
    D. Apply a cervical collar

21. The proper procedure for disabling a vehicles battery system is:

    A. Disconnect the positive side first
    B. Cutting the positive line first
    C. Disconnect the negative side first
    D. Cut the negative side first

22. The physician orders too high a dose of a drug. What would you do?

    A. Give it anyway
    B. Tell the physician that you think the dose is too high
    C. Give the dose you know to be right
    D. Ask to speak to another physician

23. Tissue anoxia from diminished blood flow, caused by occlusion or narrowing of the artery to the tissue is a definition of:

    A. Atherosclerosis
    B. Necrosis
    C. Ischemia
    D. Thrombus

24. What is a major concern when dealing with a patient with organophosphate poisoning?

    A. CVA or neurological effects
    B. Life threatening dysrhythmial
    C. Explosion or fire hazard
    D. Exposure of rescuers to the poison

25. Which of the following is the most common medical problem with neonates?

    A. Electrolyte imbalance
    B. Hypothermia
    C. Hypovolemia
    D. Hypoglycemia

26. What does the actual drawing of the QRS complex on an EKG tracing show?

    A. Impulse travel through the atrioventricular junction
    B. Only ventricular repolarization
    C. Ventricular depolarization and atrial repolarization
    D. Ventricular repolarization and atrial depolarization

27. A telemetry system in which both voice and an EKG an be transmitted from the field to the hospital at the same time is an example of a:

    A. Multiplex system
    B. Convex system
    C. Simplex system
    D. Duplex system

28. One breathing pattern is characterized by periods of apnea, followed by periods in which respirations first increase then decrease in both depth and frequency. This pattern is called:

    A. Diaphragmatic respiration
    B. Apneustic respiration
    C. Cheyne-Stokes breathing
    D. Central neurogenic hyperventilation

29. What is the most common cause for convulsions in children 6 months to 6 years of age?

    A. Epilepsy
    B. Head trauma
    C. Hypoglycemia
    D. Febrile illness

30. A common treatment for trycyclic antidepressant overdose is:

    A. Sodium bicarbonate
    B. Magnesium sulfate
    C. Calcium chloride
    D. Lidocaine

31. As you approach the scene of an auto accident, which of the following steps should be completed before all others?

    A. Making sure the scene is safe
    B. Stabilization of the vehicle
    C. Stabilization of the patient
    D. Packaging of the patient

32. If the heart rate increases, but the stroke volume remains the same, what happens to the cardiac output?

     A. Decreases
     B. Remains the same
     C. Increases
     D. The stroke volume must increase also

33. All of the following are acceptable indications for administration of intravenous nitroglycerine except:

     A. Myocardial infarction
     B. Ventricular dysrhythmias
     C. Pain relief in unstable angina
     D. Pulmonary edema

34. The permission to provide care obtained from the patient after the nature and the risks of care are explained is called:

     A. Implied consent
     B. Informed consent
     C. The patient's right to know
     D. The duty to act

35. A 50 year old female who is a mother of three complains of right upper quadrant pain with nausea and vomiting, you suspect:

     A. Ruptured uterus
     B. Ectopic pregnancy
     C. Cholecystitis
     D. Hepatitis

36. After a normal delivery of a full term infant, the one minute APGAR score is made. The baby has a pink body, but the extremities are blue. His pulse is 120 beats per minute, and he is crying lustily and jerking his arms and legs. The one minute APGAR score is:

    A. 15
    B. 3
    C. 9
    D. 11

37. The typical profile of an ambulance collision is one that occurs in the:

    A. Daytime on clear, dry roads
    B. Night involving intoxicated drivers
    C. Fall when the roads are covered with wet leaves
    D. Winter while driving on snow and ice

38. In the _____many of the first volunteer rescue squads were organized in Roanoke, Virginia and along the New Jersey coast:

    A. 1950's
    B. 1920's
    C. 1930's
    D. 1940's

39. Potential benefits of on-line (direct) medical control include:

    A. Quicker access to definitive care
    B. Treatment orders that are more efficient
    C. The ability to obtain real time direction and orders
    D. Interfacing with a physician for more effective treatment

40. _____is the study of occurrence of disease:

    A. Epidemiology
    B. Morbidity
    C. Incidence
    D. Mortality

41. Which of the following is associated with an overall lower risk of death?

    A. Occupation in health care
    B. Residing in warmer climates
    C. Residing in colder climates
    D. Higher education attainment

42. The range of duties and skills a paramedic is expected to perform when necessary is called the:

    A. Negligence per se
    B. Scope of practice
    C. Delegated authority
    D. Protocol

43. The engulfing of liquid droplets is called:

    A. Exocytosis
    B. Phagocytosis
    C. Endocytosis
    D. Pinocytosis

44. The tension exerted on cell size caused by water movement across the cell membrane is referred to as:

   A. Colloid pressure
   B. Osmotic pressure
   C. Oncotic pressure
   D. Tonicity

45. Examples of disease that are more prevalent in males include:

   A. Parkinson's disease
   B. Rheumatoid arthritis
   C. Osteoporosis
   D. Breast cancer

46. A drug that joins partially with a receptor and prevents a reaction is call a (an):

   A. Agonist
   B. Partial antagonist
   C. Partial agonist
   D. Antagonist

47. Which of the following metric units is the largest prefix?

   A. Kilo
   B. Milli
   C. Deka
   D. Centi

48. You were just given a report on a patient and were told that the patient's temperature is 101.6 f. What is the patient's temperature in Celsius?

  A. 41.3
  B. 40.1
  C. 39.5
  D. 38.7

49. You are running a micro drip at one drop every two seconds. How many minutes will it take for a 50 ml bag of normal saline to run out completely?

  A. 50 min.
  B. 100 min.
  C. 25 min.
  D. 40 min.

50. Who should decide if EMT-B's are trained to use the LMA?

  A. The Medical director
  B. The training officer
  C. The International Resuscitation Committee
  D. The American Heart Association

51. At what age do the anterior fontanelles usually close on an infant?

  A. 2 years
  B. 3 – 6 months
  C. 9 – 18 months
  D. 6 – 9 months

52. You are assessing an infant that is 11 months old. You do not know the child's exact weight, but you can estimate that the child's weight is _____ its birth weight:

    A. Six times
    B. Five times
    C. Double
    D. Triple

53. The rate of respiration can be increased by any of the following factors, except:

    A. Hypertension
    B. Fever
    C. Drugs
    D. Acidosis

54. The pons has the secondary control center of respirations called the _____ center:

    A. Secondary
    B. Chemoreceptor
    C. Apneustic
    D. Backup

55. Which of the following is a contraindication for the use of a demand valve?

    A. Dyspnea
    B. Apnea
    C. Tachypnea
    D. Bradypnea

56. Putting on gloves and wearing a face mask and a gown are examples of:

   A. Patient contact priorities
   B. Components of a physical examination
   C. Taking standard precautions
   D. Performing an assessment

57. In the pre-hospital setting, the most common site to auscultate for bruits is over the:

   A. Brachial artery
   B. Carotid artery
   C. Abdomen
   D. Upper chest

58. The components of forming a general impression of the patient include all of the following, except determining the patient's :

   A. Health history
   B. Level of distress
   C. Gender
   D. Approximate age

59. The purpose of the initial assessment is to _____and find and manage any life threatening conditions:

   A. Assess for potential hazards
   B. Obtain patient consent
   C. Determine the patient's mental status
   D. Obtain baseline vital signs

60. When a patient has both a medical and trauma problem, which is treated first?

    A. Medical control decides
    B. Trauma
    C. Medical
    D. Any threat to life

61. Where is severe trauma stabilized?

    A. In the Operating room
    B. On the scene
    C. In the Ambulance
    D. In the Emergency department

62. The difference between dizziness and vertigo is that:

    A. Vertigo is not associated with nausea
    B. Vertigo is a vestibular disorder
    C. There is no difference
    D. Only adults experience vertigo

63. Which of the following is not a key aspect of the ongoing assessment?

    A. Time constraints
    B. Manpower
    C. Scene size- up
    D. Trending

64. The p02 measures oxygenation at sea level and should normally run
     greater than _____ mmHg.

     A. 70
     B. 30
     C. 50
     D. 80

65. Cigarette smoke inhibits the normal:

     A. Movement of mucus via bronchial cilia out of the lungs
     B. Hormonal response in postmenopausal women
     C. Response to breathe in teens
     D. Immune response in the elderly

66. The layer of the heart that lines the chambers of the heart and is
     continuous with the intima is the:

     A. Visceral pericardium
     B. Endocardium
     C. Epicardium
     D. Myocardium

67. The coronary sinus, a portion of the coronary circulation, is a large
     vein that opens into the:

     A. Vena cava
     B. Right atrium
     C. Aorta
     D. Left ventricle

68. A _____is a solution containing real antibodies to bind with a specific pathogen:

    A. Shot
    B. Vaccine
    C. Serum
    D. Titer

69. Phenytoin (Dilantin) and carbamazapine (Tegretol) are both medications used to treat:

    A. Seizures
    B. Myocardial Infarction
    C. Gout
    D. Angina Pectris

70. You are treating a 230 pound patient. The medical control physician orders you to administer medications based upon the patient's metric weight. What converted weight would you use?

    A. 76 kg
    B. 500 kg
    C. 300 kg
    D. 105 kg

71. Side effects such as dilation of the pupils, dryness of the mouth, and flushing of the skin, may occur after the IV administration of:

    A. Atropine
    B. Isuprel
    C. Dopamine
    D. Morphine

72. The intramuscular injection needle is usually administered at a
    _____degree angle:

    A. 25
    B. 60
    C. 90
    D. 45

73. Which of the following is not normally a suicide risk factor?

    A. Homosexuality
    B. Married persons
    C. Living alone
    D. Previous attempt

74. In general, the toxins of aquatic animals can be neutralized with:

    A. Epinephrine
    B. Cold
    C. Heat
    D. Lidocaine

75. Which of the following lowers blood sugar by facilitating glucose
    transport into cells and by encouraging glycogen production?

    A. Thyroxine
    B. Insulin
    C. Oxytocin
    D. Glucagon

76. Penetrating trauma is dangerous when it involves the neck because it may cause problems with:

    A. The cervical spine
    B. Severe hemorrhage
    C. The airway
    D. All of the above

77. Patients with pelvic inflammatory disease often complain of which of the following?

    A. Itching upon urination
    B. Diffuse lower abdominal pain
    C. Severe vaginal bleeding
    D. Tearing pain in the uterus

78. What are the blood vessels in the umbilical cord?

    A. Two arteries and one vein
    B. One artery and one vein
    C. One artery and two veins
    D. Two arteries and two veins

79. Children's aspirin is in which of the following classes of medications?

    A. Benzodiazepines
    B. Salicylates
    C. Tricyclics
    D. Acetaminophen

80. Your patient is an 88 year old female with extreme difficulty breathing, apprehension, diaphoresis and cyanosis. Assessment findings include elevated blood pressure and pulse. Rhonchi and rales are heard on auscultation. There is no chest pain. What condition should you treat this patient for?

    A. Left sided failure secondary to a Myocardio infarction
    B. Right sided heart failure
    C. Cardiogenic shock, secondary to a Myocardio infarction
    D. Dissecting aortic aneurysm

81. A 82 year old female trips and falls while walking to her bedroom. Physical findings include a lateral rotation of the left foot and knee. The patient has tenderness and a protrusion in the left groin area. You suspect:

    A. Posterior dislocation of the hip
    B. Distal femur fracture
    C. Anterior dislocation of the hip
    D. Ischeal tuberosity fracture

82. An endotracheal tube that has been advanced too far is prone to enter which of the following structures?

    A. Right main stem bronchus
    B. Esophagus
    C. Left main stem bronchus
    D. Trachea

83. Asymmetrical movement during respiration typically suggests which condition?

    A. Brain damage
    B. Hemothorax
    C. COPD
    D. Flail chest

84. You should attempt to remove foreign material from a patient's airway with forceps only in which situation?

    A. You have tried but failed to suction the airway
    B. You are able to visualize the obstruction directly
    C. You are unable to insert an endotracheal tube
    D. You do not have access to laryngoscopy equipment

85. A 35 year old female complains of diffuse lower abdominal pain, vaginal discharge, and low grade fever. Which of the following conditions best describes the patient's presentation?

    A. Ovarian cyst
    B. Pelvic inflammatory disease
    C. Ectopic pregnancy
    D. Kidney infection

86. An adult patient has burns covering his head and upper back. Using the rule of nines, this patient's burns cover what percentage of his body surface area?

    A. 36%
    B. 9%
    C. 18%
    D. 27%

87. The wheezing associated with left sided heart failure results from:

    A. Fluid in the lungs
    B. Chest wall expansion
    C. Chest muscle tightness
    D. Chronic bronchitis

88. All of the following are identified as toxic industrial chemicals except:

    A. Chlorine
    B. Hydrogen cyanide
    C. Sarin
    D. Anhydrous ammonia

89. Soman, tabun, sarin, and VX agents are all classified as _____.

    A. Incapacitating agents
    B. Choking agents
    C. Blister agents
    D. Nerve agents

90. HIV infection may be transmitted by all of the following body secretions except:

    A. Feces
    B. Blood
    C. Semen
    D. Vaginal secretions

91. A person who has blood type _____ is called the universal recipient.

    A. B
    B. A
    C. AB
    D. O

92. The symbol or abbreviation for potassium is:

    A. K+
    B. P-
    C. P+
    D. PO+

93. Fluid with osmotic pressure equal to normal body fluid is called:

    A. Hypotonic
    B. Homeotonic
    C. Isotonic
    D. Hypertonic

94. The fetal circulation changes to the normal circulation:

    A. With expulsion from the vaginal canal
    B. With the first respiration
    C. Due to drying of the body
    D. With exposure to a cool environment

95. The white area of the eye is called the:

    A. Conjunctiva
    B. Iris
    C. Sclera
    D. Cornea

96. The inner ear is susceptible to injuries from:

     A. Basalar skull fracture
     B. Blast trauma
     C. Diving trauma
     D. Any of the above

97. The nipple line body surface region is sensed by the:

     A. Fourth thoracic nerve
     B. Fourth cranial nerve
     C. Fifth lumbar nerve
     D. Third cervical nerve

98. Adequate perfusion is dependent on which of the following elements?

     A. The container (blood vessels)
     B. The pump (heart)
     C. The volume (blood)
     D. All of the above

99. The specific type of shock due to a severe reaction to an allergen is called:

     A. Hypovolemic
     B. Metabolic
     C. Anaphylactic
     D. Psychogenic

100. Which of the following is the thin membrane covering the sclera?

    A. Renal capsule
    B. Bowman's capsule
    C. Conjunctiva
    D. Glisson's capsule

101. Which of the following statement is true with regard to synchronized cardioversion?

    A. It is not as safe as unsynchronized cardioversion
    B. It is indicated for unstable PSVT
    C. It is faster than unsynchronized cardioversion
    D. It is indicated for pulseless ventricular tachycardia

102. Which of the following will not cause sinus bradycardia?

    A. Sleep
    B. Digoxin
    C. Isoproterenol
    D. Increased vagal tone

103. Total failure of the mitral valve would eventually lead to:

    A. Dyspnea
    B. Distended neck veins
    C. Pulmonary edema
    D. All of the above

104. The three stages of a stress response include:

    A. Tiredness, drunkenness, craziness
    B. Alarm, resistance, exhaustion
    C. Alarm, rescue, contain
    D. Rescue, race, relief

105. The purpose in starting an IV in a patient with chest pain is to:

    A. For blood administration
    B. Have a medication port
    C. Have access to administer fluids
    D. All of the above

106. When injury to the myocardium affects the automaticity and the heart is no longer able to produce the normal pace or rhythm, the treatment of choice in the pre-hospital setting is:

    A. Defibrillation
    B. External pacing
    C. Atropine
    D. Isoproternol

107. When cardiac muscle tissue gets an altered amount of potassium, the effect on the heart is a (an):

    A. Fast heart rate
    B. Increased force of contraction
    C. Slow heart rate
    D. Decreased force of contraction

108. The major difference between a stroke and a TIA is that:

    A. The stroke is considered a warning sign
    B. A TIA always precedes a stroke
    C. The TIA has no lasting effect
    D. A stroke always precedes a TIA

109. Which of the following glands is responsible for the secretion of anti-diuretic hormone?

    A. Pituitary
    B. Ovaries
    C. Testes
    D. Thymus

110. During the pre-dive surface phase of a dive, any of the following potential problems may occur, except:

    A. Near drowning
    B. Motion sickness
    C. Air embolism
    D. Hyperventilation

111. All of the following are types of classifications of psychiatric disorders, except:

    A. Substance related
    B. Adolescence
    C. Anxiety
    D. Mood

112.   During fetal development, red blood cells are produced in the:

   A.   Umbilical cord
   B.   Spleen
   C.   Lungs
   D.   Kidneys

113.   The majority of the blood cells are formed in the:

   A.   Bone marrow
   B.   Spleen
   C.   Lungs
   D.   Liver

114.   When a patient has a low number of WBCs, this is called:

   A.   Anemia
   B.   Leukemia
   C.   Leukopenia
   D.   Leukocytosis

115.   Cells without intracellular granules are called:

   A.   Monocytes
   B.   Hemocells
   C.   Neutrophils
   D.   Exudates

116.   The base of the uterus is called the:

   A.   Cervix
   B.   Endometrium
   C.   Fundus
   D.   Uterine cavity

117. The hormone responsible for restoring and preparing the uterus for pregnancy after menses is:

   A. Pitocin
   B. Progesterone
   C. Estrogen
   D. Oxytocin

118. When does the "Golden Hour" for the trauma victim patient begin?

   A. Imediately after the injury is sustained
   B. Immediately upon arriving at the Emergency department
   C. When the first responder arrives at the patient's side
   D. When the paramedic arrives on the scene

119. Cervical spine injuries are most common with what type of collision?

   A. Frontal
   B. Secondary
   C. Rear end
   D. Primary

120. Intravenous volume expanders that have the same tonicity as plasma are_____ solutions.

   A. Synthetic
   B. Hypotonic
   C. Isotonic
   D. Hypertonic

121.   Fibroblasts, macrophages and MAST cells are located in the:

    A.   Deep fascia
    B.   Dermis
    C.   Epidermus
    D.   Superficial facia

122.   The "P's" of compartment syndrome, include all of the following, except:

    A.   Palpation
    B.   Pulselessness
    C.   Pain
    D.   Paresis

123.   What is a funiculus?

    A.   The center of the reticulospinal tract
    B.   A group of nerve fibers with a similar function
    C.   A ligament found in the spinal column
    D.   The key to the corticospinal tract

124.   Which of the following is the most significant finding of a speech impairment for a patient in the emergency setting?

    A.   Chronic slurred speech
    B.   Hearing loss with an acute onset
    C.   Hearing loss with a slow onste
    D.   A possible associated psychiatric disorder

125. Which of the following is considered non-invasive ventilation?

    A. PEEP
    B. ETT
    C. EOA
    D. BiPAP

ANSWER SHEET 8

1.   C- Black, yellow, red
2.   A- In a position of comfort
3.   B- Attempt to keep the patient aware of the time, place, person and situation
4.   B- Replace and cover eviscerating organs
5.   D- Both B and C
6.   C- A cloth uniform
7.   B- Transport the patient to the hospital
8.   A- Hyperextension of the neck
9.   C- The psychotic patient is not in touch with reality
10.   D- Stridor
11.   B- Perform fundal massage
12.   B- GI hemorrhage
13.   A- Obstruction of blood flow
14.   C- Lowest priority
15.   A- 200 joules
16.   B- Nitronox is a short acting agent that is administered via inhalation
17.   B- Hypertension
18.   D- Holding the radio in a vertical position
19.   B- Have suspected spinal injury
20.   A- Apply manual stabilization
21.   C- Disconnect the negative side first
22.   B- Tell the physician that you think the dose is too high
23.   C- Ischemia
24.   D- Exposure of rescuers to the poison
25.   A- Electrolyte imbalance
26.   C- Ventricular depolarization and atrial depolarization
27.   A- Multiplex system
28.   C- Cheyne-Stokes breathing
29.   D- Febrile illness
30.   A- Sodium bicarbonate

31. A- Making sure the scene is safe
32. C- Increases
33. B- Ventricular dysrhythmias
34. B- Informed consent
35. C- Cholecystitis
36. C- 9
37. A- Daytime on clear, dry roads
38. B- 1920's
39. C- The ability to obtain real time direction and orders
40. A- Epidemiology
41. D- Higher education attainment
42. B- Scope of practice
43. B- Phagocytosis
44. D- Tonicity
45. A- Parkinson's disease
46. D- Antagonist
47. A- Kilo
48. D- 38.7
49. B- 100 min.
50. A- The Medical director

ANSWER SHEET 8

51. C- 9 – 18 months
52. D- Triple
53. A- Hypertension
54. C- Apneustic
55. B- Apnea
56. C– Taking standard precautions
57. B- Carotid artery
58. A- Health history
59. C- Determine the patient's mental status
60. D- Any threat to life
61. A- In the Operating room
62. B- Vertigo is a vestibular disorder
63. C- Scene size-up
64. D- 80
65. A- Movement of mucus via bronchial cilia out of the lungs
66. B- Endocardium
67. B- Right atrium
68. C- Serum
69. A- Seizures
70. D- 105 kg
71. A- Atropine
72. C- 90
73. B- Married persons
74. C- Heat
75. B- Insulin
76. D- All of the above
77. B- Diffuse lower abdominal pain
78. A- Two arteries and one vein
79. B- Salicylates
80. A- Left sided failure secondary to a Myocardio infarction
81. C- Anterior dislocation of the hip

82. A- Right main stem bronchus
83. D- Flail chest
84. B- You are able to visualize the obstruction directly
85. B- Pelvic inflammatory disease
86. C- 18 %
87. A- Fluid in the lungs
88. C- Sarin
89. D- Nerveagents
90. A- Feces
91. C- AB
92. A- K+
93. C- Isotonic
94. B- With the first respiration
95. C- Sclera
96. D- Any of the above
97. A- Fourth thoracic nerve
98. D- All of the above
99. C- Anaphylactic
100. C- Conjunctiva

ANSWER SHEET 8

101. B- It is indicated for unstable PSVT
102. C- Isoproterenol
103. D- All of the above
104. B- Alarm, resistance, exhaustion
105. D- All of the above
106. B- External
107. C- Slow hear rate
108. C- The TIA has no lasting effect
109. A – Pituitary
110. C- Air embolism
111. B-Adolescence
112. B- Spleen
113. A- Bone marrow
114. C- Leukopenia
115. A- Monocytes
116. C- Fluids]
117. B- Progesterone
118. A- Immediately after the injury is sustained
119. C- Rear end
120. C- Isotonic
121. B- Dermis
122. A- Palpation
123. B- A group of nerve fibers with a similar function
124. B- Hearing loss with and acute onset
125. D- BiPAP

www.ingramcontent.com/pod-product-compliance
Lightning Source LLC
Chambersburg PA
CBHW031824170526
45157CB00001B/175